AF324709

CHANGING PATTERNS OF INTERNATIONAL COOPERATION IN SPACE

by
Joan Johnson-Freese
Director, Center for Space Policy and Law
University of Central Florida

with Foreword by
David C. Webb

ORBIT BOOK COMPANY

MALABAR, FLORIDA
1990

Original Edition 1990

Printed and Published by
ORBIT BOOK COMPANY, INC.
2005 Township Road
Malabar, Florida 32950

Library of Congress Cataloging-in-Publication Data

Johnson-Freese, Joan.
    Changing patterns of international cooperation in space / by
Joan Johnson-Freese.
       p.   cm—(Orbit, a foundation series)
    ISBN 0-89464-022-4 (alk. paper)
    1. Outer space—International cooperation. 2. Outer space—
International cooperation—Government policy—United States.
I. Title.
QB500.26.I58J64   1990
33.9′4—dc20   89-377110
                                          CIP

10 9 8 7 6 5 4 3 2

Contents

Series editor
Edwin F. Strother, Ph.D.

Foreword

We must be grateful to Dr. Joan Johnson-Freese for holding up a mirror in which we can examine our record in the field of international cooperation in space. We may not like what we see, but then, perhaps, we will be more careful about how we act in the future. For if this work has one central message, it is that the United States lacks a long-term policy regarding international cooperation in space.

NASA has more international *agreements* (1,200) than most nations; but these are bilateral agreements around specific issues. NASA is only one agency, and a small one at that, of the U.S. government. The fact that it has negotiated so many agreements on behalf of the United States is an interesting issue in itself (one that may usefully bear further analysis by political scientists) but it does not, nor cannot, create a *national policy*.

It is unfortunately only too evident from reading this work that we, as a nation, have been and still are very reluctant to commit ourselves to any particular long-term policy in the international space field. We apparently prefer to muddle along, developing strategies (if that is what they may be called) around specific issues. The result is a hodge-podge of decisions, or non-decisions, that upon review look like the proverbial crazy quilt.

Is that any different than any other area of U.S. political life, the reader may rightly inquire. The answer, of course, is that it is not. But the costs, in terms of creating our own competition and alienating our good friends, may be higher in the space arena than in most others, as Dr. Johnson-Freese correctly infers. Herein lies the value of the mirror analogy; for it is quite evident that despite all the bilateral agreements, and all the efforts of many well-meaning people inside and outside NASA, our nation's record in the international space field leaves much to be desired.

This kind of knowledge should now spark a long overdue examination of the haphazard manner in which we deal—or do not deal—with the manifold issues arising from a rapidly technologizing world, one in which the United States is playing a declining role.

Dr. Johnson-Freese performs a major service in reminding us that time does not stand still. History should teach us that the nation or nations that fail to adapt to changing circumstances, particularly in meeting new challenges, quickly fade into obscurity. The somber message of this work is that although we were the big boy on the block at the beginning of the space age, others have not stopped growing. By joining together, they will displace us if we do not continue to earn our position of leadership and learn to accept that they have much to offer in what should be a global effort to explore and develop space. Rather than attempting to fight the inevitable, we should actively initiate cooperative space endeavors with other nations in order to further this great evolution in human affairs.

Dr. Johnson-Freese is to be congratulated for tackling a work of this scope. It should be required reading for everyone concerned with U.S. science and technology. The need for the development of a new cooperative attitude should not be restricted to space affairs alone. The entire issue of how we, as human beings, continue to develop and implement technological innovations is today a matter for global concern. We now know with greater certainty than any previous generation that there can be considerable costs incurred by the uncontrolled introduction of technologies without any measurement or understanding of their possible adverse effects on the environment.

We must start somewhere, however. Space exploration and development represent our best opportunity. What we learn in that area will stand us in good stead in comprehending and responding to even greater demands in other areas of science and technology. There is no better time to start than 1992, the five-hundredth anniversary of Columbus's discovery of the New World and the first International Space Year celebrating humankind's entry into the newer world of space.

DAVID C. WEBB, Ph.D.

Acknowledgments

This book is the result of an interest in international cooperation in space first prompted by Dr. Hermann Strub, head of Aerospace Programs at the West German Ministry for Research and Technology. Acting as host for a visit by Dr. Strub to the University of Central Florida campus in 1983 was a more fortunate and pleasant experience than I could ever have imagined. I am grateful to him for his continued encouragement to pursue space-related research.

In writing this book a great many people have been invaluable in many different capacities. My husband John dealt with my anxieties—not an easy task. Joyce Lilie, chair of the Political Science Department, encouraged me to actually take on this project and reminded me to stick with it if I tended to get distracted—as academics often do. Stuart Lilie saved me when I blew a computer disk and lost many pages of my draft revisions.

I have conducted numerous interviews, both formal and informal, in researching this book, and tried to cite as many as possible of those interviews in the text of the manuscript. I am grateful and wish to thank all of those people who gave so generously of their time. Jim Brice, my student assistant at UCF who has a bright future in the space program, was a big help with research and manuscript preparation; he was further assisted by Glen Lockwood. Mary Roberts at Krieger Publishing has been very patient with me, and always supportive. And finally, David Webb, Burton Edelson, and Ian Pryke all took time from their busy schedules to read draft manuscripts and provide insightful comments. Their willingness to take on this task is indicative of the professionalism, dedication, and cooperative spirit with which each of them and the vast majority of people in the space field approach their work. Responsibility for the book content, however, remains solely my own.

Finally, this book is for JB who, if he chooses to, really will have the chance to go to space someday.

Introduction

In its relatively short history, "space" has accumulated a remarkable history of cooperative efforts. The U.S. Space Act of 1958, which created NASA as a civilian agency to take the lead in pioneering U.S. space activities, includes a provision that NASA "may engage in a program of international cooperation in work done pursuant to this Act, and in the peaceful application of the results thereof. . . ." To its credit, NASA has taken that proposition very seriously, as evidenced by the numerous international agreements concluded and carried out over the years.

Industrial European countries realized long ago that in order for them to be able to compete internationally, they would have to cooperate internationally also. The Canadian space program has also been built on cooperation: their space activities really beginning with the joint U.S.–Canadian Alouette satellite in 1962. Today, Japan, China, the Soviet Union, and many other countries are actively seeking and participating in cooperative ventures.

Clearly, however, not all areas are ripe or even appropriate for joint efforts. Specifically, areas of national security and areas where proprietary rights over technology could be compromised have traditionally been deemed off limits. By engaging in cooperation, one party may in fact find itself helping another party which will eventually become a competitor in a particular area. Hence there is a somewhat contradictory nature to cooperation. Therefore, the key to successful cooperation—where all parties see themselves as benefiting—is to identify areas where you can cooperate to increase your own capabilities without compromising technology or national security.

The basic thesis of this book is that in fact over the years certain factors have become evident as impacting cooperation. These factors become relevant, however, only after national security and technology transfer considerations have been made. One may in fact posit a conditional relationship between the factors which will influence the degree, nature, and type of international cooperation that is possible: **If outside the realm of technology transfer or national security, then attitude, potential benefit, and political environment are all factors affecting cooperation in space ventures.** These factors are, of course, interdependent as well.

National security considerations have been and continue to be critical in the decision-making process concerning cooperation on space ventures, particularly in the Soviet Union and the United States. Within the United States, there have often been times when the administration in office has had a different agenda than Congress and scientists where cooperation has been concerned. For example, U.S. scientists James Van Allen and Lloyd Berkner were among those responsible for the initiation, planning and promotion of the International Geophysical Year (IGY) for 1957–58, with the support of the U.S. government, to coordinate worldwide high-altitude research. Included in the scientists agenda for IGY was the launch of a U.S. satellite—just as the Soviets had announced similar intentions. However, unbeknownst to U.S. scientists, the Eisenhower administration also had other, higher priority items on their space agenda. The administration's primary concern at the time was to establish the legality of satellite overflight, for the purposes of its planned reconnaissance program, which was basically stated in the 1955 document NSC-5520. This was probably not the first, but certainly an early example, of civilian space projects being secondary to the military space agenda; a pattern which has been consistent over thirty years. Within this book, that relationship should be taken as a given, as it is outside of the scope of this examination to consider the important issue of the awkward and forced relationship between the military and NASA.

Within those factors to be considered, political environment refers to general conditions under which negotiations are undertaken—or not undertaken—as the case may be. Apollo-Soyuz was possible in 1975, whereas it probably would not have been in 1962 during the cold war, or 1983 when President Ronald Reagan subscribed to the Evil Empire view of the Soviets. Political environment is the variable which should be considered first in any analysis, as the predeterminant within which the others operate.

Attitude may be defined at either or both a macro or micro level. That is, individual attitudes of key personnel have often been critical as to whether or not an international venture has gone forward or been stifled. The attitude of an agency administration as a whole can be equally important in determining the political atmosphere in which individuals operate, regardless of how internationally minded they are or are not.

Potential benefit can take on a political, economic, scientific, or technological meaning. The degree of importance given to a potential benefit often varies according to attitude. President John F. Kennedy, for example, gave a great deal of weight to the political/public relations potential of space, in terms of national and international image. The Reagan administration, on the other hand, for the first time stressed the commercial potential of space. The Europeans have often stated that their initial reason for wanting to work with the United States in space was to learn and build their own technology base. Scientists want to work together to increase the potential scientific rewards, which basically means data return, of a space science mission. Cooperation occurs when parties feel they have something to gain by it. This attitude should not be taken as nefarious, but rather as pragmatic.

This book examines the history of international cooperation in space in a framework of three periods: The United States Leads the Way, 1958–69; An Increase in Cooperative Activity, 1970–1984; and Multiple Actors Give Rise to Choices, 1985–current. Within that context, each of the three variables cited is examined as to whether each individually was conducive to, or an inhibitor of, cooperation. This is done by case study examinations or a review of activities during the time frame under consideration. Subsequently, propositions are generated regarding how to maximize opportunities for successful cooperation in the future.

There are two key questions which must be examined for each of the historical periods of international space cooperation to be examined: what were the motivations behind cooperative ventures, and how were cooperative ventures carried out. Like the technology used in space, the answers to these motivational and implementation questions become increasingly more complex with time.

Finally, cognizance should be taken of the problems involved, particularly in Part III of the book, with trying to analyze a constantly changing field. It is rather like trying to narrate a baseball game in progress: things change before you can even analyze them. Therefore, rather than being primarily concerned with a journalistic reporting of current events, every attempt is made to identify policy trends indicative from current events. With the rapidity of change that occurs, however, it is not always possible to provide neat answers to policy questions presently under consideration.

PART I

The United States Leads the Way: 1958–1969

In a 1987 speech, Reimar Luest, director-general (D-G) of the European Space Agency (ESA), characterized the metamorphosis of the relationship between the National Aeronautic and Space Administration (NASA) and ESA as having gone through three stages: "1. the Tutorship of the United States to Europe. 2. Europe as the Junior Partner of the United States and 3. Partnership and Competition between Europe and the United States."[1] This characterization can easily be expanded to apply to NASA international relationships generally, and to provide a framework for examining the history of international cooperation—and competition—in space.

The total time period involved in an examination of international space activities is a relatively short one, even by U.S. standards. It can be said to officially commence in 1958 with the U.S. National Aeronautics and Space Act, which created NASA and included in its objectives:

> Cooperation by the United States with other nations and groups of nations in work done pursuant to this Act and in the peaceful application of the results thereof.[2]

NASA actively embraced this objective, for a variety of reasons, until approximately the end of the Apollo era, 1969. Those initial reasons, or motivations for cooperation, are examined (Chapter 1), as a starting point to trace the evolution of cooperation. After Apollo, it is not a change of fundamental attitudes that results in a change in the environment, but the beginning of a change in the environment that will subsequently result in a changing of attitudes.

The United States was not alone in the quest for space. The Soviets were pursuing an equally aggressive path to space as the Americans (Chapter 2). Other countries, looking primarily at the potential political and economic importance of space, also began preparing themselves for the future (Chapter 3). Their efforts began slowly, but served the intended purpose of getting these countries on the learning curve.

1958–1969 is characterized as the "Golden Age" of the U.S. Space Program, and not without due cause. After the October 1957 launch of Sputnik, the United States suffered a slow and sometimes embarrassing start playing catch-up with the Soviet Union. Having Project Vanguard, the first U.S. attempt at launching a civilian satellite, blow up on the launch pad in full view of the press and consequently the American public in December 1957, was certainly not the entry into the space age that had been hoped for. But the United States eventually rallied and responded to the challenges presented by the Soviet Union in a way that staggered imaginations worldwide. The challenge was first addressed in an all-out fashion by the Kennedy administration, which recognized the political, economic, and strategic importance of space, and was willing to stand by the president's rhetorical commitments, and by the technology itself, with a can-do spirit that took the United States to the Moon.

In launching Sputnik, the Soviet Union threw down a gauntlet that American scientists, engineers, industry—and even politicians while circumstances made it worth their while—responded to in an unprecedented manner. The United States entered the civil space age in a race, one not formalized until 1961 with President Kennedy's declaration of the goal to send a man to the Moon by the end of the decade, but nevertheless one where there were two key opponents, and several peripheral players anxious to build their capabilities to the point of being active participants. At the same time, international infrastructures were beginning to be developed to deal with the multitude of new legal and organizational issues brought forth by the space age (Chapter 4). Because of the self-imposed isolation of the Soviets, the United States was able to lead in the development of these also. Hence, the course followed by the Soviet Union inherently made the United States the only choice as a "professor" in virtually all major aspects of the tutorial relationship that followed with these peripheral players.

Director-General Luest quotes a "top-level participant" in the U.S. space program as saying "when resources abound and opportunities are plentiful, a cooperative attitude abounds . . . When the resources and opportunities shrink, . . . altruism takes a back seat and . . . scientists take a more selfish view of cooperation."[3] The United States was a bountiful land of opportunity where space was concerned during the Golden Age, hence it could afford to be generous.

Endnotes

1. Reimar Luest, "The Cooperation of Europe and the United States in Space," The Fulbright 40th Anniversary Lecture, 6 April 1987, Washington, D.C., p. 5. Accordingly to Luest, these phases are not unique to space either, but in fact have been evidenced in other areas of science also, such as high-energy and elementary particle physics or plasma physics. The lecture is reprinted as "Cooperation between Europe and the United States in Space" in the *ESA Bulletin*, May 1987, pp. 98–104.

2. Public Law 85-568, National Aeronautics and Space Act of 1958, July 29, 1958.

3. Luest, "The Cooperation of Europe and the United States in Space," p. 10.

Chapter 1

Initial U.S. Motivations for Cooperation

Why Cooperate?

Initially the "why" of international cooperation for the United States was relatively simple and pragmatic. As Congressional Research Service (CRS) space policy analyst Marcia Smith put it, NASA had both a "genuine desire to involve other countries in exploring the new frontier of space" and "required worldwide tracking locations and wanted to generate a climate in which other countries would be predisposed to allow tracking sites on their territory."[1] Actually, there was the necessity of establishing four tracking programs: (1) mini-track, a north-south network throughout the Western hemisphere for scientific satellites, (2) the deep-space network, (3) the manned space flight ground stations, and (4) the Baker-Nunn (named for the camera design used) tracking stations, for a Smithsonian astrophysics program.

In those programs, the usual agreement was that the host country provided land and NASA provided equipment. In most instances the station was operated by nationals, with the exception of the station director position. The entire operation resulted in generating employment, training programs, operations money (which could run into millions of dollars), and other general benefits for the area. Because these sites generated desirable and tangible benefits for the host countries, they were sometimes more difficult to get out of than they were to get in to. At least once, when NASA was ready to leave a site because it was no longer needed, the issue raised very serious political difficulties between the host country and the United States. It was also the case that twice NASA was asked to leave its tracking sites by host countries. In both instances, involving African countries, there were internal political reasons rather than direct disputes with NASA.

During the early U.S. communication satellite programs, having a local ground station became something of a status symbol in other countries. European countries competed with each other for the privilege of being at the receiving end of satellite communications. This was front page news, and several countries were willing to build the ground stations at their own expense. In the case of what was to be the first transmission, some other countries even attempted to steal the glory from Britain, which was to receive the first transmission, by picking up the test pattern and claiming first reception. So NASA's pragmatic need for tracking sites was not ill-received elsewhere. Besides the economic benefits to be had, in many cases it was the only way countries could enter the glamorous and exciting new space age.

During the Kennedy administration, NASA was in the enviable position of being under the direct purview and protection of the White House. This protection had a very positive affect on the Apollo program, allowing it to successfully reach fruition in a phenomenally short time period. Unfortunately, this situation also got NASA into the habit of expecting a particular working environment—one where money and support is plentiful—which was to prove a very short-lived phenomenon. Subsequently, operating in a politically competitive environment has been a difficult situation for NASA to adjust to in some instances. The protection of the Kennedy administration basically allowed NASA the luxury of simple motivations where early international cooperation was concerned. But the motivations of the United States, and consequently NASA, have evolved into an ever increasingly complex and interrelated set of objectives. This sometimes has resulted in mixed signals being given and contradictory goals being sought by various government agencies within the United States, each with a voice in the space project or program in question. Basically, however, international program objectives have and continue to fall into three general areas: scientific and technical, economic, and political.

Scientific and Technical Objectives

Originally the scientific and technical objectives of space cooperation flowed mainly from the United States to other countries. The October 1987 report of the NASA Task Force on International Relations in Space pointed out that "measured in terms of technology acquisition, the returns to the United States have not been significant to date."[2] Returns are likely to increase in the future, however, because of the scientific and technological maturing of other space programs. The maturing of these programs has also led to an environment where it is sometimes difficult for the United States to meet some other of its objectives.

Cooperation increases available brainpower. The United States clearly recognized that scientists and engineers from other countries could make a valuable contribution to our space program. This premise dates back to Operation Paperclip in which U.S. troops sought German rocket expert Wernher von Braun and his colleagues to bring them to the United States.

The recognition of the capabilities of Canadian radioastronomers was one of the reasons the Alouette project became the first NASA cooperative satellite program, launched in September 1962. Alouette was a basic research project to investigate the properties of the ionosphere. The findings from that project had practical implications concerning the transmission of radio signals. Therefore, it provided a good model for other cooperative programs in that it was strictly scientific in nature, but had practical implications to make it interesting to a larger community. Alouette then led to the International Satellite for Ionospheric Studies (ISIS) project in 1969 and 1971.

Cooperation allowed the United States to guide or shape the development of foreign space programs, to be compatible—and perhaps even complementary—to the U.S. program. In the past, the primary implementation method for this objective was through offering other countries attractive opportunities to work with the United States. Also, persuading countries to commit funds to a cooperative project concurrently limits the amount of funds which they then have available for a potentially competitive project.

Economic Objectives

Early economic objectives were of a very general nature. Cooperation was seen as a way to promote international economic development. Cooperation came also to be recognized as a way to improve the U.S. balance of trade by creating new global markets for U.S. communications and aerospace products.

Political Objectives

Although the political objectives of international cooperation may appear the most indirect, in reality they have been a primary objective, particularly in the early years. In fact, the NASA Task Force on International Cooperation in Space states that ". . . international cooperation in space from the outset has been motivated primarily by foreign policy imperatives."[3]

Cooperation created a positive image and prestige for the United States and reinforced an image of openness to contrast with the Soviet Union during the cold war years. Kennedy saw this as particularly important in reference to third world countries which equated technological leadership with political leadership.

A fourth category of motivations for cooperation, regulation, should perhaps also be included, although it is difficult to discern whether in fact it was an objective or a benefit stemming from cooperation. Through maintaining its leadership role for so long in the international space community, the United States was generally able to shape the currently existing regulatory measures in which space activity is conducted. Regulation involves the development of a legal framework for general space activity and international mechanisms for managing limited space resources, like geostationary orbits. Unilateral national actions could have been not only impractical but also potentially disruptive to U.S. space activities.

Cooperation in general makes it easier to allocate scarce or finite resources, such as geostationary orbits. Even prior to the introduction of satellites, organizations such as the United Nations and the International Telecommunications Union were involved with coordinating and regulating certain activities of global dominion. The International Telecommunication Satellite Organization (INTELSAT) extended those premises, reflecting the early recognized need to coordinate telecommunication satellite activities.

U.S. Guidelines for Cooperation

Equally important to the "why's" of cooperation, have been the "how's." This basically refers to a discussion of means and ends, and whether or not the means used by NASA to carry out the ends have been appropriate and have been modified as the goals or emphasis on particular goals have changed.

The guidelines used by NASA in cooperative projects have been:[4]

- Designation by each participating government of a government agency for the negotiation and supervision of joint efforts.
- Conduct of projects and activities having scientific validity and mutual interest.
- Agreement upon specific projects rather than generalized programs.
- Acceptance of financial responsibility by each participating agency for its own contributions to joint projects.
- Provision for the widest and most practicable dissemination of the results of cooperative activities.

Each of these points had a specific purpose and reason

behind it, which worked very well in the context of the 1960s and even into the 1970s. Because the United States was dominant in space, it could dictate terms of cooperation to other countries, which they were more than willing to accept in order to gain entrance to the space program. For the political benefits, the United States wanted to work with other countries on *science* projects of mutual interest and clear scientific value where we could benefit from their personnel and involve others without technology transfer problems. Each side had to have the technological and managerial capability to carry out their own work. The importance of these criteria is evidenced in the fact that technical agreements were settled before political commitments were made. Agreements at that time were primarily bilateral and always *project specific* so that U.S. terms could be clearly defined, including the provision for *no exchange of funds*. This provision was politically necessary not only for the United States, so that U.S. dollars would not be leaving the country and hence the project lose requisite political support, but for the same reason in other countries as well. Because nearly all of the early projects were scientific, the results could then be *widely distributed*, again creating good-will and furthering the learning curve for the benefit of all, the United States included.

In order to evaluate whether the means used to achieve the desired ends of cooperation have been appropriate, a criterion must first be stated upon which a judgment will be made. The October 1987 report of the NASA Task Force on International Relations in Space defined cooperation as ". . . a generic term used loosely. It covers a wide range of modalities and intensity, all having in common parties working or acting together for mutual benefit on agreed topics."[5] It goes on to discuss that in the past, most NASA cooperative projects originally were simple, consisting primarily of NASA flying a foreign experiment on a U.S. spacecraft. The task force points out that all forms of cooperation should be intended as "bargains in which each side seeks something it values. Cooperation is intended to create a win-win situation although it does not always work out that way."[6] Therefore, it can be said that cooperation should result in both sides being satisfied that they have gained something of value equal to, or greater, than what they have put into the project.

Conclusion

The cold war, East versus West political environment that was predominant throughout much of the early years of space cooperation made identification of potential partners in cooperative space ventures relatively simple: cooperation was both sought and allowed between the United States and its Western bloc allies, and with countries that the United States wanted to attract to the Western alliance. As long as there was little or no technology transfer and national security interests were not compromised, cooperation offered extensive political benefits: a strengthened Western alliance and the admiration, with the hope of subsequent loyalty, of the third world. At the same time, cooperation also increased political and technological pressure on the Soviet Union.

Terms of cooperation during this period were basically dictated by the United States, and accepted by others. Initially, these terms of cooperation were dictated as much by a realistic assessment of the capabilities of potential cooperative partners as by specific design. Under this mode of operation the United States developed several expectations concerning cooperation, which later would prove difficult to alter when the environment for cooperation changed. The United States began to direct cooperative projects, both as a function of leadership and of technological capabilities. Conversely, in some cases this created a mindset that equated leadership with project direction; if one did not direct a project—or all projects—then clearly one could not be a leader. Further, the United States developed an image of its potential partners as technologically immature and needing the United States. Once created, images have proved hard to dispel. And finally, although the United States originally placed political benefits high on its list of goals to be achieved through cooperative space ventures, they soon became expected and consequently taken for granted. This will pose serious problems later, when "leadership" again becomes an issue in the U.S. space program, but under a new set of environmental guidelines.

U.S. motivations for involving other countries in its early space ventures included aspects of both pragmatism and altruism. Space technology could have been guarded much the same way as nuclear technology has been; but it was not. To its credit, the United States made the deliberate decision to involve other countries in its space program. Typically, problems commence only after those benefiting from cooperation with the United States progress along the learning curve, which is inherent in and in fact a goal of cooperation, to a point which results in competition.

Endnotes

1. Smith, Marcia S., "America's International Space Activities," *Society*, January/February 1984, p. 18.
2. NASA Advisory Council, Task Force on International Relations in Space," "International Space Policy for the 1990's and Beyond," October 12, 1987, p. 18.
3. Ibid.
4. NASA, Division of International Affairs, *26 Years of NASA International Programs*, January 1, 1984, p. 2.
5. NASA Advisory Council, Task Force on International Relations in Space, "International Space Policy for the 1990's and Beyond," October 12, 1987, p. 17.
6. Ibid.

Chapter 2

Behind Closed Doors: The Soviet Space Program

Cold War Politics

Until recently, the Soviet space program has been closed to the West. The Soviet space program—like the U.S. space program—has been dominated by the military. Unlike the United States, however, the Soviet space program initially made no attempt to separate a portion of their space efforts for civilian, open activity. Therefore until the mid-1980s, when the Soviets deliberately allocated, and separated as much as possible, a portion of their space program for civilian activities, cooperation with the West was politically impossible. Further, the development of both the U.S. and Soviet space programs took place against the backdrop of the cold war, when military competition was preponderant. Cooperation between the United States and the Soviet Union was a rhetorical goal more than anything else, and subsequently both space programs managed to develop in virtual isolation of each other.[1]

During the 1950s and 1960s, the United States was making a very deliberate effort to establish the reputation of the U.S. space program as being open, cooperative, peacefully oriented, and for the good of all mankind as opposed to the closed, secretive, militaristic Soviet space program. Of course, the United States had a closed, secret, military space program also, but it was intentionally kept in the background. But maintaining this peaceful, open image placed the United States in somewhat of a dilemma. Historian Walter McDougall characterizes the resultant American space diplomacy in the 1960s as one of "benign hypocrisy." McDougall says:

> Then there was the U.S.S.R. The whole American enterprise rested on the assumption that the Soviets were leading in space. If so, how could they be denied equal influence over space law and international management? If the United States did "leapfrog" the Soviets, or if real cooperation put an end to the race, then how could NASA continue to justify big budgets? If the United States then cut back on space budgets, it would only provide the Soviets and other commercially minded Europeans and Japanese with a "window of opportunity" to catch up in selected fields and to smash U.S. hegemony over time. . . .

These facts of life informed an American space diplomacy with Soviets and Europeans alike of cooperation in science and competition in engineering; of agreements in areas where the United States was safely dominant and advocacy of laissez-faire in areas where there was still a race. It all suggests a "benign hypocrisy," a hard-headed approach not quite up to American rhetoric. Yet U.S. statesmen would have been derelict in their national duties had they acted differently.[2]

There were two areas of cooperation with the Soviets which were discussed in the 1950s and 1960s: arms control and space science. Ironically, it was in the area of arms control where cooperation was on more than just a rhetorical level, simply because it was in the national interest of both countries.

Arms Control

The primary motivations behind both the original Soviet and U.S. ventures into space were military. The United States wanted to develop a satellite reconnaissance system to watch Soviet activities, while the Soviets sought an accurate and reliable delivery system for nuclear weapons. Therefore, both superpowers had a vested interest in establishing and maintaining the military ground rules of that domain to their benefit.

Early on, it was decided by U.S. strategists that bombs in orbit would be of little military value, for a variety of reasons. A primary factor at the time was that land based ICBMs were cheaper, more accurate, and more controllable. The Soviets, on the other hand, did not have the technology to make space weapons a viable option, yet it was exactly the scenario of Soviet bombs in orbit which was played up as a reality and subsequently turned into a crisis by the technically uninformed press after the launch of Sputnik. Neither superpower really wanted to see this happen. Space weapons were viewed as an invitation to an expensive arms race. Therefore, for virtually no cost, what better way for both the United States and the Soviet Union to boost their respective images internationally, than to cooperate in the renunciation of something that neither wanted anyway. The three arms control agreements that were the result of this attitude were the Partial Test Ban Treaty, the UN Principles on Outer Space, and a ban on orbital weapons of mass destruction.

Space Science

The second area of rhetorically supported cooperation during the 1950s and 1960s between the United States and the Soviet Union was space science. But here a distinction must not only be drawn between what was said and what was done, but also between what was said and what was really desired.

The motivations for the United States wanting to engage in cooperative space activities have already been discussed, and public opinion and idealism ranked high in those motivations. As the country with the most advanced space technology in almost every area though, the United States then had to contend with the dilemma of potential technology giveaways as a consequence of cooperation, which was certainly not in its best interest. The problem was particularly acute concerning the Soviets, with whom the United States was in a clear competition, even after the initiation of detente.

What emerged was a U.S. policy which stressed looking for areas of basic space science that could be investigated together, and clearly avoiding areas of applied, technology driven exploration. Within this context, it also was considered imperative that the United States be perceived as the willing, cooperative partner being stonewalled by the hard-nosed, militaristic Soviets. As the Soviet program really was hard-nosed and allowed for little or no cooperation with the West in any area, the Kennedy administration knew it was safe in repeatedly offering to work with the Soviets, even on projects where the United States clearly did not want them—like the Apollo program. In a September 20, 1963, speech at the United Nations, President Kennedy proposed the possibility of cooperation with the Soviets in space exploration, including manned lunar flight. Even earlier, at the June 1961 Vienna summit, Kennedy brought up the idea of going to the Moon together, which of course Soviet Premier Nikita Khrushchev summarily rejected as Kennedy knew he would.

Following a pattern which would become increasingly typical over the years, cooperation, when it did occur, was on space science projects. This was true even prior to Sputnik. One of the earliest attempts at cooperative scientific activity was the International Geophysical Year (IGY). The IGY actually extended for a period of eighteen months, from July 1957 to December 1958. At the suggestion of University of Iowa scientist Dr. James Van Allen to the International Council of Scientific Unions (ICSU), IGY was designated to encourage international efforts in studying the physical environment of the Earth, the oceans, the atmosphere, and outer space. One of the areas focused on for the IGY was the launching of an artificial satellite.

Even during the IGY, which the Soviets rhetorically supported, they showed the utmost reluctance to engage in any meaningful data exchange. This was not due to problems between the scientists, but due to politics. Again, the lack of separation between the military and the science aspects of the Soviet space program virtually tied the hands of scientists wanting to work together, and meant that the Soviets worked primarily in isolation behind closed doors. The Soviets had announced in 1956 that they intended to launch a satellite as part of IGY. Few Americans believed them, although Wernher von Braun had been warning Washington of this likelihood for almost two years. Subsequently, when the Soviets launched Sputnik 1 in October 1957 as part of the IGY, it caught the U.S. public and Congress by surprise.[3] The entire effort is interesting in that it illustrates a trend which follows rather consistently throughout space history, that of international space scientists trying to work together in spite of political differences.

The closed nature of the Soviet space program also extended to international organizations. When INTELSAT was founded in 1964, the Soviet Union and its bloc allies refused to join. Their objections were focused on U.S./COMSAT management, the use of U.S. technology and to the system of weighted voting whereby voting power was determined by a country's use, in percentage terms, of the system. Initially, Soviet need for satellite usage was estimated at only 2 to 3 percent, as opposed to 50 percent by the United States. Therefore, in 1968 the Soviet Union, Poland, Czechoslovakia, East Germany, Hungary, Romania, Bulgaria, Mongolia, and Cuba proposed an alternative system, which was formally agreed to in 1971 and called Intersputnik. The Intersputnik system initially was based on use of Molniya satellites. These are nongeosynchronous satellites, which made it difficult and expensive for use with INTELSAT Earth stations, which work with fixed geosynchronous satellites.[4] This augmented the political difficulties of cooperation with technical incompatibilities.

Space continued to be dominated by an attitude of competition between the United States and the Soviet Union in the 1960s. The Soviets managed to capture numerous places in the record books with space firsts in the early years of the decade, including Yuri Gagarin's flight as the first man into space in April 1961; launching the first woman (Valentia Tereshkova) into space, June 1963; the first three man flight in October 1964, and the first space walk (Alexei Leonov) in 1965. The intensity of the competition is illustrated by the fact that the Soviets accomplished all these "firsts" with very little technical change or advancement in their hardware. At Khrushchev's insistence, they merely modified their Vostok and Voskhod capsules to enable them to complete the desired mission quickly and beat the United States to the record books. In one case, safety equipment was merely removed to make room for extra seats to enable them to "progress" from a one-person to three-person capsule. This also illustrates how much of the Soviet program during the early years was based on illusion rather than technical advancements. Carrying through Khrushchev's demands for "firsts" drove L. A. Voskresensky, one of Soviet "Chief Designer" Sergei P. Korolev's assistants, to a nervous breakdown.[5]

An agreement was signed between NASA and the Soviet Academy of Sciences in 1962 stipulating coordinated national efforts in the fields of meteorology, geomagnetism, and satellite communications experimentation. Generally known as the Dryden-Blagonravov agreement, after NASA's Hugh Dryden and the Soviet Academy of Sciences' Anatoly Blagonravov as the primary negotiators, the results were disappointing. Part of the problem can be attributed to the then poor level of Soviet technical capabilities for processing data. But the general Soviet emphasis on secrecy, exacerbated by Cold War politics, was the primary inhibitor.

The Soviet penchant for secrecy is part of their basic conservative nature and has made them extremely cautious about exposing themselves and their possible faults to others. This attitude was particularly heightened by the success of the United States with the Apollo program.[5] Recognizing that the problems they were having with their heavyweight launch vehicle would inevitably lead to the United States reaching the Moon first resulted in the Soviets redirecting their efforts away from the Moon and into low Earth orbit, long duration manned flights, where they have been very successful. But the spectacular achievements of Apollo created serious psychological hesitancy in the Soviets about exposing themselves to further potentially embarrassing situations, especially in an area where they had originally been viewed as the worldwide leader. It would take many years of building successful launches and creating their own niche in low Earth orbit manned flights before the Soviets would be ready to "go public" with their space program.

Conclusion

During the 1950s and much of the 1960s, competition was the watchword between the United States and the Soviet Union. Therefore, as the antithesis of compe-

tition, cooperation was impossible. In fact, cooperation among countries of the Western world was at least in part motivated by the desire to more effectively compete, technically, economically and politically, with the Soviet bloc.

But whereas the political environment made cooperation in space projects a virtual impossibility, the premise that cooperation occurs in self-serving areas held true. The political benefits of concluding particular arms control agreements were such that the competitive attitude of both U.S. and Soviet leaders was overcome. It would not be until much later that the other blocks to cooperation, on both sides, could be dismantled.

Endnotes

1. For other works on early space history, see: *U.S.–Soviet Cooperation in Space* (Washington, D.C.: U.S. Congress, Office of Technology Assessment, OTA-TM-STI—27 July 1985; Arnold Frutkin, *International Cooperation in Space* (Englewood Cliffs, NJ: Prentice-Hall, 1965); Walter Sullivan, *Assault on the Unknown: The International Geophysical Year* (New York: McGraw-Hill, 1961); Homer E. Newell, *Beyond the Atmosphere; Early Years in Space Science* (Washington, D.C.: National Aeronautics and Space Administration, 1980).
2. Walter McDougall, . . . *The Heavens and the Earth* (New York: Basic Books, Inc., 1985), p. 345.
3. Several interesting theories have been posited as to whether or not the Eisenhower administration was really surprised. Walter McDougall has said, "The satellite race may have been lost by the United States more or less deliberately—to help clear the legal path for strategic reconnaissance by satellite." "Sputnik, the Space Race, and the Cold War," *Bulletin of the Atomic Scientists*, May 1985, p. 20.
4. Geosynchronous satellites are of a more sophisticated nature in that they are in a fixed position relative to an Earth tracking station. The nongeosynchronous satellites could be picked up by the fixed trackers, but major satellite modifications and a computer to track the satellite would be needed. There would also need to be several more stations around the globe to keep in contact with the satellites, similar to NASA's ground-based tracking for manned orbital missions. Obviously then, economical, political, and technical factors would come into play.
5. McDougall, . . . *The Heavens and the Earth*, p. 292.
6. Nicholas Daniloff examines the Soviet attitudes toward space in his book *The Kremlin and the Cosmos* (New York: Alfred A. Knopf, 1972), particularly Chapters 5 and 6.

Chapter 3

Early Peripheral Players in Space

Why Space?

In the 1950s and 1960s space meant technology, technology meant industrialization, and industrialization meant economic growth. It was that simple, and nobody wanted to be left behind. This fear of being left behind is evidenced in 1960s literature by such European analysts as Jean-Jacques Servan-Schreiber, and later in the 1970s by Roger Williams.[1] According to space historian Walter McDougall,

> De Gaulle's *certaine idée* of the future of France encompassed more than *la gloire* and distrust of 'the Anglo-Saxons.' It depended above all on technological self-sufficiency, both military and economic. Between 1959 and 1963, France's R&D spending quadrupled, with much of the new money going to two new aerospace agencies—the Centre National d'Etudes Spatiales (CNES) and the Office National d'Etudes et de Recherches Aerospatiales (ONERA)—and to the Society pour l'Etude et la Realisation d'Engins Ballistiques (SEREB), which would be responsible for the initial development of both military and civilian rockets.[2]

France was not the only country that embraced this desire for technological independence.

The only two countries with the technology capable of reaching space were the United States and the Soviet Union. The Soviets had made it clear they had no intention of working with Western nations; that made the United States the only game in town for those looking to learn from those who already had the expertise. Working with NASA gave foreign scientists, engineers, and managers invaluable experience and allowed them to build their own learning curve, not to mention invaluable flight opportunities. Making these offers of cooperation allowed the United States to assert its leadership in space.

Europe

Motivations

Although a so-called "European" perspective can be stated, it should also be remembered that encompassed within that perspective are some decidedly different viewpoints. Generally speaking, however, the drive for technology, industrial application, and subsequent economic benefit was and remains the key motivation for European activities in space.

Many European countries have national space programs as well as being involved in cooperative ventures. Beyond fear of technical domination by the United States, original European national interests in space have traditionally been somewhat divided, and this has been reflected in their national space programs. France and West Germany have been the two major European space powers, but for differing reasons. The Germans have been inclined toward pursuing an active space program for scientific research, research which they hope could have industrial and hence economic payoffs in the long run.[3] France, however, was originally more concerned with the military side of a space program. As stated, the French wanted launch capability independent of the United States. That they were able to link it with civilian payoffs was at first seen as somewhat of a side benefit, though an important one. Much later, the primary civilian "payoff" came through the commercial competitiveness of the Ariane launcher. An analogy may be drawn between the Ariane program and the French nuclear power program, which originally was focused on military applications but resulted in Europe's largest civilian nuclear power program.

France, through the Societe pour l'Etude et la Realisation d'Engines Ballistiques (SEREB), began with development of their "precious stone" series of boosters: the Agate, Topaze, Rubis, etc. In 1965, France became the third country in the world with launch capability with the liftoff of their Diamant-A rocket from Hammaguir in the Algerian Sahara, putting the Asterix satellite into orbit. France did not see itself as competing with the United States. Rather, France believed that Europe as a whole was destined to eventually move away from military, economic, and technological dependence on the United States. The goal was for France to be the European space leader.

According to McDougall, the reason that France succeeded was due in part to the ambivalence of Great Britain.[4] Upon reflection, perhaps British ambivalence was merely a reaction to the frustration of the bureaucratic quagmire which has—and continues to—plague their space program. The responsibility for space has been assigned to no less than nine ministries, sometimes simultaneously. After World War II, Britain took the lead in European missile development by working on prototypes of their Blue Streak and Black Knight missiles, but cancelled their military missile program after the advent of Sputnik, opting instead for reliance on the United States.[5]

With the efforts being made in the 1960s toward European unity in terms of the European Economic Community, cooperative work in space ventures certainly made sense also. History has proven that projects and programs too ambitious—in terms of both cost and manpower—for any one European country alone, might well be managed as a cooperative effort. Initially, however, the drive for technology did not easily convert into a space program. In fact, finding a raison d'être was perhaps the most serious challenge that faced early European efforts at cooperation.

In the late 1960s the objectives of European space policy became more precise, including:

- Give the European scientific community the possibility of remaining in the forefront of space research;
- Help Europe master advanced technologies, through meeting the challenging requirements of space engineering;
- Promote the use of space technologies for a number of applications (telecommunications, meteorology, remote sensing) for the benefit of Europe's economic wealth.[6]

But much like in the United States, European motivations, as reflected in their objectives, have become increasingly more complex.

Basically, European motivations fall into the same three categories as U.S. motivations: scientific and technical, economic, and political. Reimar Luest has outlined them to include:[7]

- Scientific and Technical
  —Brainpower
  —Coordination
  —Sharing resources
  —Exchanging opportunities
  —Stimulating the exchange of advanced technology
- Economic
  —Strengthening Western industrial capacity
- Political

—Strengthening the transatlantic partnership
—Increasing European unity

The operational priorities accorded to each element are critical and will change over time and in reaction to events.

Organizations

According to Arnold Frutkin, NASA director of International Affairs during the formative years, European motivation for the formation of a separate space infrastructure can be rather simply stated. In his 1965 book he said:

> An adequate outlet for European *scientific* interests may have existed in the opportunities for cooperation with NASA in the United States. But the Europeans want to master the *technology* of space activity. This accounts for the support given to the establishment of ESRO and ELDO . . .[8]

The European Space Research Organization (ESRO)[9] and the European Launcher Development Organization (ELDO), whose conventions went into force in 1964, served as Europe's initial efforts at coordinating space activities. The former, which focused on satellite and sounding rocket programs, was far more successful than the latter, though it was not without problems. Satellite launches did not begin until 1967, and then all aboard NASA boosters. All of the satellites launched were of a scientific nature only, as opposed to commercial, which fit in well with NASA philosophy, but did little to encourage investment of further European capital. Furthermore, there was bitter disagreement among the ESRO governments concerning the disproportionate distribution of industrial return from satellite contracts. France, for example, contributed only 19 percent of ESRO's budget, but received 37 percent of all ESRO outlays between 1965 and 1967. Although this is a tribute to French technology policy and the strength of their political will, it also caused some hard feelings.

The purpose of ELDO was straightforward: build a European rocket to break European dependence on U.S. launch technology. The Europa-1 was to be the vehicle to achieve that goal. The idea was to have each participating country build part of this rocket, which would be launched from (Woomera) Australia. Although individual components were of quality design—the Blue Streak first stage from Britain, the Coralie second stage from France, the German Astris for the third stage, with Italy supplying the test satellite—the combined product simply did not work. By 1969 ELDO had managed to launch nothing, but go more than 300 percent over budget in the attempt.

The failure of the Europa was a microcosm of the failure of the early European space efforts. The creation of two agencies to handle intimately linked space tasks, although it was intended that they would work closely together, allowed typical organizational problems to be exacerbated. Rather than cooperate, the two agencies decided basically to ignore one another, even though for a time they resided in the same building, 114 Charles de Gaulle, Neuilly, respectively on the sixth and seventh floors.[10]

Britain had insisted that the Blue Streak be used as the first stage of the Europa rocket, making such use a precondition for their participation in the program. When the Europa program failed, Britain withdrew and France was then left with the opportunity to take over as the dominant European country to lead the next attempt at building an independent European launch capability.

Cooperation

Whether collectively or on an individual country basis, and regardless of motivation, most early European space activity was part of a NASA effort: either by flying experiments on NASA missions, cooperative sounding rocket missions, ground-based experiments, cooperative balloon and airborne projects, solar energy projects, scientific and technical exchanges, or eventually through cooperative spacecraft projects.

ESA Director-General Reimar Luest has stated that:

> Of course, every single European state and later Europe as a whole, gladly accepted the help offered by the U.S., through NASA for their first steps into space, since for the countries initially approached, such as the U.K., France, the Federal Republic of Germany, Italy, it was the most efficient way to build up capabilities and capacity in space science and technology and obtain in a reasonably short time scientific results from space missions.[11]

Former NASA Deputy Director of International Relations James Morrison succinctly characterized the European rationale: "Industrial European countries long ago realized that in order for them to compete internationally, they would have to cooperate internationally."[12] Their early recognition of that premise, based on need and experience in other areas, would later prove advantageous.

Europe worked with the United States in many of its early activities. NASA launched the first two ESRO scientific satellites, ESRO II (IRIS) on 17 May 1968, and ESRO I (Aurorae) on October 3, 1968, free of charge with Scout rockets. NASA worked with individual European countries also. By the end of 1969, nine European spacecraft, from France, West Germany, Italy, and the United Kingdom, had been launched by NASA on a cooperative basis.

Canada

Canada was the fourth country, after the Soviet Union, the United States, and Britain,[13] to have a satellite in orbit. This was the Alouette I, launched in 1962 as the second NASA cooperative spacecraft project, and was the first satellite to be built by a nation other than the United States or the Soviet Union. NASA had entered into negotiations with the Canadian Defence Research Board (DRB) in November 1958, very soon after its own inception. An attitude of amiability between the organizations is reflected in a letter from A. H. Zimmerman, chairman of the DRB, to NASA Administrator T. Keith Glennan:

> Action is being taken to arrange advance blanket clearances for laboratory visits on the part of DRB personnel. We will welcome similar arrangements on your part.
>
> I am pleased that DRB and NASA are able to undertake this important project on a co-operative basis.[14]

The Memorandum of Understanding (MOU) later signed by both agencies in 1963 carefully delineated the responsibilities of both agencies to avoid technology transfer, but the cooperative spirit of the project was reflected throughout.

This basic science project to investigate properties of the atmosphere had practical implications for the Canadians to advance along the telecommunications learning curve. That this was indeed their wise and practical intent is evidenced from their achievements today.

> Telesat Canada, the world's first domestic communications satellite company, was established as a crown corporation by an act of the Canadian parliament in 1969 to provide commercial satellite service across the country. The first satellite for Telesat Canada, ANIK A1, was launched in 1972. Between 1972 and 1984, eight additional ANIK satellites were sent into orbit.[15]

Once Canada had demonstrated their mastery of the science, their motivation evolved from scientific knowledge to applied technology.

Canada's desire for satellite technology can best be understood in terms of geography and population. First, the vast geographic area which Canada encompasses and its relatively small population create unique problems; communications technology would reduce isolation and also increase national productivity. Also, the Canadian geography creates unique opportunities: Canada is the best, or among the best, locations in the world to observe certain space science phenomena like auroras and magnetic storms. As Canadians made the decision not to develop their own launch capability, that inherently made it necessary that they base their space program on international cooperation.

Japan

Although it is sometimes said that Japan was a relatively late entry into the aerospace field because of prohibitions imposed by the United States at the end of World War II, there have been Japanese scientists interested in space since the 1950s. Scientists at the Institute of Space and Aeronautical Science (ISAS) at the University of Tokyo began in about 1955 to launch sounding rockets for the International Geophysical Year 1957–58. The university has maintained a strong interest in space science since that time and has been responsible for major Japanese capabilities in that area.

From a national viewpoint, however, the motivation behind the Japanese space program is unquestionably economic. Although initially reliant on the United States for technology, Japan's long-term objective has always been to attain both technological and political autonomy in space—much like Europe and China. Space technology is part of Japan's more general goal of establishing itself and strengthening its competitive position within the world market for advanced technology.

The Japanese government established the National Space Development Center in 1964, which became the National Space Development Agency (NASDA) in 1969. NASDA has primary responsibility for projects involving the practical applications of space developments. It is supervised jointly by the Japanese Science and Technology Agency, the Ministry of Posts and Telecommunications, and the Ministry of Transport. ISAS has retained responsibility for scientific satellites, the MU launch vehicle, and promotion of space science activities. ISAS is under the supervision of the Ministry for Education.

China

In 1956 the Chinese developed a twelve year plan designed to incorporate rocket and jet engine technology into key state projects. The first rocket research institute was founded that year, with Vice-Chairman Zhou Enlai serving as its leader. In 1958, Chairman Mao Zedong proposed that China build satellites. That the Chinese were serious in their efforts is evidenced by the firing of their first indigenous sounding rocket in February 1960. Later that year, they launched a mockup of a ballistic missile.

Originally, the motivation behind the Chinese efforts was primarily political and military, specifically building a strategic deterrent capability against the Soviet Union. Organizationally, the Chinese Academy of Science established a space committee in 1963 which developed plans for a satellite launch in July 1965.

India

As in China, Indian scientists were involved in space research even before the launch of Sputnik. Indian space research was under the guidance of a cosmic ray physicist, Vikram Sarabhai, at his Physical Research

Laboratory in Ahmedabad. Sarabhai is considered the father of the Indian space program. Coming from a wealthy and philanthropic family of textile manufacturers that has been likened to the Rockefellers, Sarabhai was able to pursue his interest in space and encourage others as well. His interest coincided with the belief of Indian Prime Minister Jawaharlal Nehru, that India needed technology to be strong. Therefore, although India obviously could use the financial resources devoted to space to deal with more immediate needs, the expected returns from space over time have been considered important enough to warrant consistently spending a portion of their limited funds on space activities.

Originally, the Indian Department of Atomic Energy (DAE) held jurisdiction over space activities. In 1962 Nehru created the Indian National Committee for Space Research (INCOSPAR) within the DAE, with Sarabhai as chairman. Through Sarabhai's efforts in that position, a UN launching range for sounding rockets was established at Thumba, in southern India. Equipment for the facility was donated by the United States, the U.S.S.R., and France. Operations began in 1963 at the site, and through the Scientific and Technical Committee of the United Nations Committee on the Peaceful Uses of Outer Space (COPUOS), it received official UN sponsorship in 1965. The Indian Space Research Organization (ISRO) was established in 1969, also under the auspices of DAE, and eventually became part of INCOSPAR.

Conclusion

For those countries which had both the foresight and the funds, the 1950s and 1960s were a time of preparing themselves for a future which would most certainly involve space. The exact nature of the role that space would play in international economics and politics was still somewhat unclear, but its potential importance was not. Countries which had any aspirations of playing an active role in a future world where technology would act as a driver in many fields, realized that they could not afford anything less than an early commitment, even where in some cases that commitment meant sacrifice elsewhere.

The political environment of the fifties and sixties encouraged cooperation within the Western bloc and offered a plethora of political benefits for all parties. The practical benefits of cooperation on space projects accrued primarily to those countries at the low end of the learning curve. The fact that cooperation initially offered the United States little in terms of practical return was of little consequence, as that was not a primary concern at the time.

The wisdom of early posturing would soon become apparent. Those countries which became involved in space early have been able to form a second tier of space powers, some of which are now actively attempting to join the United States, the Soviet Union, and Europe on the first tier.[16] Also, countries which became involved early would later become leaders within their own groups or regions. In many cases, the decisions made during this first era of space cooperation set the pace for future patterns of cooperation.

Endnotes

1. J. J. Servan-Schreiber, *The American Challenge* (New York: Avon Books, 1967); Roger Williams, *European Technology: The Politics of Collaboration* (London: Croom Helm, Ltd., 1973).

2. Walter McDougall, "The Scramble for Space," *The Wilson Quarterly*, Autumn 1980, p. 72.

3. See: Michael Feazel, "Germany Cites Commercial Fallout as Justification for U.S. Station Involvement." *Commercial Space* (Spring 1985): pp. 47–54.

4. Ibid., p. 73.

5. The Black Arrow program was briefly revived in the late sixties and placed one satellite in orbit in 1971 before the program was again cancelled.

6. *Twenty Years of European Cooperation in Space 1964–1984*, European Space Agency, Paris, 1984, p. 246.

7. Reimar Luest, "Cooperation Between Europe and the United States in Space," Presented to the Association of American Universities, the Fulbright 40th Anniversary Lecture, 6 April 1987, Washington, D.C., p. 20.

8. Arnold W. Frutkin, *International Cooperation in Space* (Englewood Cliffs, N.J.: Prentice-Hall, 1965), p. 133.

9. The ten member nations of ESRO were: Belgium, Denmark, France, West Germany, Italy, the Netherlands, Spain, Sweden, Switzerland, and the United Kingdom.

10. Albert Ducrocq, "Cooperation ESA-NASA," *Air et Cosmos*, 21 January 1984, p. 37.

11. Luest, "Cooperation Between Europe and the United States in Space," p. 6.

12. Interview with James Morrison, Washington, D.C., 4 May 1988.

13. This was the Ariel, considered to be the first international satellite, launched on a Delta rocket in April 1962.

14. Letter from A. H. Zimmerman to T. Keith Glennan, 16 December 1959.

15. Ronald W. Neville, "Canada in Space," Joint American Astronautical Society/Japanese Rocket Society Symposium, December 15–19, 1985, Honolulu, Hawaii, p. 3.

16. Opinions vary, of course, as to which countries belong on which "tier" of space powers. Unquestionably, the United States and the Soviet Union would be placed on the first tier because of their vast activities and accomplishments. The Europeans are also included as a first tier power because only they can also claim experience and success in building man-rated space hardware—Spacelab.

Chapter 4

Legal Frameworks and International Organizations

Setting Guidelines

The legal frameworks built over the years concerning space activities have focused on two areas: regulation of communication satellites through international organizations and the more general regulation of what is "acceptable" activity in outer space through arms control treaties. Both of these areas have offered substantial benefits to the initiating and supporting parties. Further, in any examination of international cooperation in space, satellite communications is a key point of reference for all other activities, as over the years satellite communications has unquestionably established itself as having the best cooperative track record in the space field.

Communications Satellites

In 1945 internationally known science fiction writer Arthur C. Clarke, while then a little known engineer working for the British Post Office, wrote an article in a magazine called *Wireless World* about the idea of using space stations to link telephone services throughout the world. The concept was correct; the portion of his scenario that was incorrect was the supposition that this service would have limited use due to cost and technical difficulties. Clarke's idea later became a multibillion dollar enterprise, when independently prompted work at Bell Laboratories resulted in the first telecommunications satellite. Today, communications satellites are the only real example of space commercialization on a large scale.

In a relatively short time, an enormous new industry was created. The years between 1958 and 1964 are considered the experimental years of telecommunications satellites. During this period the TELSTAR satellite provided the first transatlantic live television broadcast. This was followed by rapid growth in operational satellites commencing with the launch of Early Bird, the first commercial communications satellite, in 1965.

INTELSAT

Regulation of communications satellites is primarily through organizations affiliated with the United Nations (UN). Membership in the International Telecommunications Satellite Organization (INTELSAT) is open to states which are members of the International Telecommunications Union (ITU), a specialized agency of the U.N. headquartered in Geneva, Switzerland, responsible for developing worldwide regulations for telecommunications. INTELSAT is an intergovernmental organization in the area of satellite communications "whose main objective is to provide, on a commercial basis, the 'space segment' for international public telecommunications services, including satellites and other equipment and services required to support the operation of the satellites."[1] Participation is through investment shares issued according to the percentage of utilization of INTELSAT's space segment by all parties.

The origins of INTELSAT[2] trace back to 1960, when advances in the United States on communications satellites reached a point when it was clear that decisions needed to be made as to how to deal with this fledgling industry. Traditionally, the United States has maintained a policy of noninvolvement with commercial research and development (R&D), theoretically at least, in keeping with the spirit of capitalism and competition. But communications satellites have been treated somewhat differently, and in a way which illustrates that innovative techniques can and should be used to encourage private sector participation in space ventures, by making it profitable.

Satellite R&D was originally conducted under the auspices of U.S. government funding and sponsorship because of the military applications satellites would offer, especially in the area of reconnaissance. Very early on, however, civilian communications satellites were recognized by the private sector as having the potential to be the wave of the future, with economic and strategic benefits to be reaped by the leader in the field. Telecommunications were experiencing phenomenal annual growth, including a large percentage of international traffic.[3] Private industry recognized that communication satellites could be an enormous money-making proposition if handled properly. Hence by 1960 companies like RCA, AT&T, GE, Hughes Aircraft, and Bell Laboratories were heavily involved with basic communication satellite research. To take this research to an operational level meant investments of a scale too large for private industry alone as, even considering potential returns, the technological and financial risks involved were still substantial.

NASA was also interested in communication satellite research, primarily from the viewpoint that satellites should be designed and developed in accordance to government interests. NASA reports to the White House pointed out that communication satellite systems would be of an inherently international nature and subsequently many countries would try to exercise jurisdictional claims. In that context, the Kennedy administration picked up on the foreign policy implications of communication satellites and the need for them to be handled in accordance with U.S. national interests. In the same speech as he made his commitment to put a man on the Moon, Kennedy requested an additional $50 million beyond what was already being spent by the government on communication satellite R&D, to conduct further research and hence ensure the earliest possible use of communications satellites on a worldwide basis. Subsequently, it was a NASA project, Syncom, which paved the way for the first commercial satellite, Early Bird, in 1965. NASA was therefore uncharacteristically involved with the R&D investment necessary to give U.S. industry the capability to build operational telecommunication satellites.[4]

As the technology leader, the United States was faced with promoting an international communications satellite system for foreign policy purposes while establishing a regime which protected U.S. leadership in the field, a point critical to the private sector. Kennedy's initial announcement was followed in July 1961 with a policy statement that there was to be a single global communication satellite system operational as soon as technically feasible. The policy clearly favored private ownership and operation of the system, but the public interests—national interest/foreign policy interests—were protected through provisions to encourage foreign participation on a nondiscriminatory basis, competitive bidding for contracts, consideration of foreign interests in the structure of ownership and control, affordable rates, and compliance with antitrust laws. There is reason to believe that the creation of a privately owned company to handle the U.S. share of the system and the point about antitrust laws were deliberate moves by the Kennedy administration to prevent

any company already in operation from gaining a worldwide monopoly in this area.

The entity created to represent U.S. interests in the international satellite system, through the Communications Satellite Act of 1962, was U.S. government-owned Communications Satellite Corporation (COMSAT). In 1964 the International Telecommunications Satellite Organization, INTELSAT, was created. As the United States operated communications satellites through COMSAT, and COMSAT worked through INTELSAT, and because the United States had the leading technology, other countries had the option of working with us, on our terms, or chance being left out of the newest innovations in the communications field. Although the terms of the agreement clearly favored the United States, most other countries still felt that their best strategy was to join the organization and try to influence future development from the inside, which later proved a correct assumption on their part. The terms that the United States initially offered were basically those that it offered in other areas of space cooperation: provide benefits to others in areas which did not negatively affect U.S. national interest as inducement for cooperation, while ensuring no technology transfer from the United States.

Originally, COMSAT Corporation was to have 61 percent of the international system, Europe 30.5 percent, with the rest to Canada, Australia, and Japan. Membership was extended to other countries as well, with the provision that the COMSAT share had to be maintained at a minimum of 50.6 percent. Management of the system was designated to the only entity that at the time was able to do the job—COMSAT. COMSAT did in fact manage INTELSAT via a contract from INTELSAT's Board of Governor's for the first six years of its operation.

Over time, an increase in INTELSAT membership led to a similar increase in pressure for restructuring of the organization to limit the influence of COMSAT. The result was transformation of a loose international consortium into an international organization grounded in two international agreements: "one concluded by the participating states and the other (the "Operating Agreement") by the states or public or private 'telecommunications entities' designated by member states."[5] Article 6 of the Operating Agreement states that ". . . each signatory shall have an investment share equal to its percentage of all utilization of the INTELSAT space segment by all signatories." As of 1988 the U.S. investment share, through COMSAT, is 26.4 percent. The next highest investment shareholder is British Telecommunications, with 14.2 percent. Therefore although COMSAT is no longer the operating agency of INTELSAT, it continues to play a large role within INTELSAT as the official U.S. designate.

The decision-making system in INTELSAT today is based on a weighted voting procedure, adjusted to the amount of investment shares each governor on the Board of Governors of INTELSAT represents, with the provision that no governor may cast more than 40 percent of the total votes. Substantive decisions require the support of at least four governors representing a minimum of two-thirds of the shares or all but three governors, regardless of the total investment share represented by the majority.

The extent and importance of INTELSAT's network can be seen from its growth. In the years since its inception, INTELSAT's membership has grown from 15 to 116 countries. The net assets of INTELSAT are approximately $1.8 billion, with 1987 revenue at $519 million from sixteen satellites in use. On a practical level, the importance of INTELSAT was already recognized and well-stated by former COMSAT Vice-President Burton Edelson in a 1977 article:

> In the decade since then (1965, when INTELSAT I was first placed in orbit) satellites have changed the pattern and pace of global communications. They have become the most obviously practical of all space benefits, providing high quality, reliable and cost-effective telephone, television and data-transmission services.[6]

In consideration of INTELSAT's accomplishments, however, it is equally important in a study of international cooperation to make special note of the precedent-setting international arrangements which have made INTELSAT a success story.

It is no small feat to bring together 116 countries of every imaginable size, economic structure, and political system to work together in a peaceful, purposeful, and profitable venture. Why it works can be traced back to the same three factors. INTELSAT has been a tool for using technology to the benefit of all involved parties. To illustrate the point, the U.S.-built satellite ANIK-A1, which was put into orbit for the Canadian organization Telesat as the first domestic satellite, grew out of the Intelsat IV satellite design. There have been consistent evolutionary patterns in satellite communications, from which all involved parties have been able to benefit.[7]

Conditions under which communication satellites exist (economic, political, and technological) are clearly different today than they were in the 1960s. Whether or not international commercial satellite communications should continue to be offered solely by INTELSAT, or whether such services should be available on a competitive basis is an issue which has been raised.[8] As a continuing model of cooperation, however, INTELSAT clearly has set an example for others to follow.

Others

Maritime communications are handled by the International Maritime Satellite Organization (INMARSAT), established in 1979. It provides communications links to more than 1,600 ships and has fifty-six countries as members, including both the United States and the U.S.S.R. INMARSAT is of particular interest and importance because it represents an area of European leadership in space activities, rather than American, and because it is the first international commercial space organization that the Soviet Union has participated in[9]—perhaps because it is not U.S. led. INMARSAT has six satellites in use (leased) and three of its own satellites on order.

Through the World Meteorological Organization (WMO), a global system of coordination and cooperation regarding data received from weather satellites has been developed. The WMO (then operating as the International Meteorological Organization, IMO) became a specialized agency of the UN in 1947 and began functioning as the WMO in 1951. Services include the World Weather Watch (WWW) to provide timely warning of storms and other weather hazards for the protection of life and property, and the World Climate Program, which studies long-term weather patterns. As weather services in virtually all countries now depend on information from satellites, cooperation is imperative and will increase as long-term climate studies progress.

EUTELSAT and ARABSAT are examples of two telecommunications organizations which operate on a regional basis. Headquartered in Riyadh, Saudi Arabia, ARABSAT has twenty-two member countries and two satellites in use. EUTELSAT has twenty-six member countries, three satellites, and is headquartered in Paris. Having once worked through ESA, EUMETSAT, with a focus on meteorological satellites, now operates on its own with a membership of twelve member countries. NORDSAT, AFSAT, PANAMSAT, and AUSSAT are regional systems currently in the planning stages.

Arms Control

The advent of man moving into outer space carried with it implications for national security. Governments became aware of these implications at approximately the same time as populations worldwide began to grasp that space offered heretofore unavailable opportunities for mankind—and against mankind. The United States pursued a determined course to assure the world that the mission of the United States in space was to explore space for peaceful purposes. It was very careful, however, not to use the word "nonmilitary." Although the difference is subtle, it is not without very important meaning. Even the interpretation of "peaceful" was the subject of debate at the United Nations during the 1960s. The smaller nations took the position that "peaceful" should be broadly defined to mean the antithesis of "military," rather than only "nonaggressive." Of course the United States took the latter position, that "peaceful" meant nonaggressive and that no weapons systems were involved,[10] to assure allowance of the continued use of photoreconnaissance satellites by the military.

Almost immediately after the launch of Sputnik in 1957, efforts were initiated to establish a means through the United Nations to deal with space issues. It was not, however, until November 1961 that the first meeting of the United Nations Committee on the Peaceful Uses of Outer Space (COPUOS) was held. This organization came about as a result of controversy concerning the militarization of space, and in anticipation of disputes that would certainly arise over conflicting interests in space. In that regard two COPUOS subcommittees were established, Legal, and Science and Technology, with a great deal of interaction between the two. Traditionally, COPUOS has operated on a consensus decision-making arrangement. This has served a useful purpose as much COPUOS work since its establishment has focused on developing treaties, or a legal regime for space.[11]

Throughout 1961–62, the United States and the Soviet Union engaged each other in verbal accusations of being the party responsible for stalling disarmament. The Eisenhower administration had put forward a proposal for general and complete disarmament (GCD), which was followed by Kennedy's "The United States Program for General and Complete Disarmament in a Peaceful World." The latter was presented to the Sixteenth General Assembly in September 1961. This was later modified in April 1962 and presented by the United States at the eighteenth nation disarmament conference in Geneva. Included in this overall disarmament package were provisions for disarmament measures in space. Both the United States and the Soviet Union took the position that steps toward disarmament in space were conditional on acceptance of GCD. As the GCD package was a political impossibility at the time, disarmament in space was thought to be unlikely also.

In March 1962, Canadian Secretary of State for Foreign Affairs Howard Green offered the U.S. delegation in Geneva an unusual proposal, which took everyone by surprise. It was a proposal for a separate ban on weapons of mass destruction in space. Although the initial U.S. reaction was negative because the proposal did not include the then standard Western demand for "inspection and control" provisions, it was given serious study in Washington.[12] The differing opinions on whether or not to pursue such a course divided as to whether the benefits of avoiding an undesirable arms race in space through separate provisions of this type outweighed the risk of bringing the issue of spy satellites into the open for possible debate, as the United States was in no way willing to have controls imposed in that arena.

Six United Nations General Assembly resolutions concerning space were adopted between 1959–1963. These were later incorporated into the Outer Space Treaty of 1967. The Outer Space Treaty was drafted by COPUOS.

> The treaty provides that outer space, including the moon and other celestial bodies, is free for exploration and use by all states and cannot be appropriated by any nation. Any exploration and use must be carried out for the benefit of all countries on a basis of equality and in accordance with international law, with due regard to interests of other states and without any harmful contamination of the environment of the celestial bodies and the earth itself. States conducting activities in outer space must, to the greatest extent feasible and practicable, disclose information about such activities and open their stations and equipment to inspection by other states.[13]

Although clearly stressing the peaceful aspects of space exploration and rejecting national claims to the Moon and other celestial bodies, the one article in the treaty dealing with the military in space was clearly an important motivation behind the treaty as a whole.

The language of the Outer Space Treaty is in places subtle and deliberately ambiguous. That is, the Moon and the celestial bodies are reserved "exclusively for peaceful purposes," which is very different from nonmilitary use.

One of the main concerns at the time the treaty was drafted was orbiting nuclear weapons. The Outer Space Treaty provides that nuclear weapons or any other kinds of weapons of mass destruction may not be placed in orbit, which means that the Earth's orbit may be used for other military purposes, particularly for launching reconnaissance satellites such as those used for compliance verification in connection with arms control agreements. Both the United States and the Soviet Union, as well as more than eighty other states, are parties to the Outer Space Treaty.

The Outer Space Treaty has also been further augmented by the Rescue of Astronauts Agreement of 1968, the Liability for Damage Caused by Objects in Space Convention of 1974, and the Agreement Governing the Activities of States on the Moon and Other Celestial Bodies of 1979. Only a limited number of states are party to the latter, and neither the United States nor the Soviet Union has signed it. Some space law specialists feel that the 1979 Moon Treaty may signal a second generation of space law,[14] as it deals with property rights in space, an area not focused on in the past. The Soviet Union, other Warsaw Pact members, Cuba, and Mongolia also adopted a 1976 Agreement of Cooperation in the Exploration and Use of Outer Space for Peaceful Purposes.

Conclusion

For the most part, the arms control measures signed in the 1960s and 1970s reflect countries agreeing not to do what they did not intend to do anyway—at least at the time. Cooperation on a selective basis was therefore a relatively low cost and high return activity designed to reap political and economic benefits. Obviously, with the advent of the Strategic Defense Initiative under the Reagan administration some of these once innocuous agreements have taken on new relevance.

With regard to communications satellites, the United States acted with aplomb and foresight in the 1960s with the creation of COMSAT to advance the technology, allow for the international utilization of the technology, and protect U.S. industrial interests. Although the exception to standard U.S. operating procedures, the success of the organization ought to give legislators pause to consider using it in other selected, appropriate high-tech fields also, as other countries have done on a regular basis. One of the advantages of being a leader in a high-tech field is being able to reap the benefits of the technology. This is not nefarious, just good business which results in profits and jobs, aspects of space activities that legislators who fund space activities can understand.

With regard to international cooperation INTELSAT—as a model for successful international space cooperation—offers valuable insight into prerequisites for making future cooperative ventures worthwhile and hence successful: (1) parties should get back returns commensurate with investment (a premise the European Space Agency has implemented) and (2) the cooperative project as a whole should yield returns greater than possible through individual efforts. The simplicity of the premises should not be mistaken for simplicity in implementation. Recognition of the premises is, however, a starting point.

Endnotes

1. Robert L. Bledsoe and Boleslaw A. Boczek, *The International Law Dictionary* (Santa Barbara, Calif.: ABC CLIO, 1987), p. 166.
2. For a complete technical description of INTELSAT operations, see: Burton I. Edelson, "Global Satellite Communications," *Scientific American*, February 1977, Vol. 236, No. 2, pp. 58–73.
3. For information on the growth of telecommunication traffic, see: Peter Brunt and Alan I. Taylor, "Telecommunications and Space," in *The Exploitation of Space*, Schwarz and Stares, eds. (London: Butterworth & Co., 1985), pp. 79–80.
4. Peter Brunt and Alan I. Taylor, "Telecommunications and Space," in *The Exploitation of Space*, Schwarz and Stares, eds. (London: Butterworth & Co., 1985), p. 82.
5. Bledsoe and Boczek, *International Law Dictionary*, p. 167.
6. Burton I. Edelson, "Global Satellite Communications," *Scientific American*, February 1977, p. 58.
7. See: Peter Brunt and Alan I. Naylor, "Telecommunications and Space," *The Exploitation of Space*, Schwarz and Stares, eds. (London: Butterworth & Co., 1985), pp. 80–82.
8. Marcellus S. Snow, "An Economic Issue in International Telecommunications: Natural Monopoly in Commercial Satellite Systems," *Economics and Technology in U.S. Space Policy*, Molly McCauley, ed. (Washington, D.C.: Resources for the Future, 1987), pp. 219–241.
9. *International Cooperation and Competition in Space Activities* (Washington, D.C.: U.S. Congress, Office of Technology Assessment, OTA-ISC-239, July 1985), p. 50, ft. 41.
10. Arnold W. Frutkin, *International Cooperation in Space* (Englewood Cliffs, N.J.: Prentice-Hall, 1965), p. 151.
11. Nathan Goldman, "Transition and Confusion in the Law of Outer Space," in *International Space Policy*, Papp and McIntyre, eds. (New York: Quorum Books, 1987), pp. 158–159.
12. See: Paul B. Stares, *The Militarization of Space: U.S. Policy, 1945–1984* (Ithaca, N.Y.: Cornell University Press, 1985), pp. 82–90.
13. Bledsoe and Boczek, *International Law Dictionary*, pp. 177–178.
14. Goldman, "Transition and Confusion in the Law of Outer Space," p. 164.

PART II

An Increase in Cooperative Activity: 1970–1984

The increased cooperation of the second period was possible because at least some nations, or groups of nations, actually had the technical capability to contribute to a project. Within Europe and for many other countries, cooperation has been and remains a means of survival. That realization prompted the formation of the European Space Agency, which gave the European countries the ability to both cooperate—and compete—with the United States and other space-faring nations in some areas. The motivations of Europe and other countries for wanting to have space programs have not been fully appreciated or understood in the United States. Until they are, the United States will be at a handicap in working with others.

At the same time the European space program was maturing, the mindset at NASA was changing, caused by changes in the domestic political environment which was reflected in the NASA budget and the low priority given to the civilian space agency (other than on an occasional rhetorical basis) by the presidential administrations of the period. Strife was becoming not only external, but internal as well, with NASA centers having to compete with each other for scarce funds. As NASA personnel were having trouble working with each other, it is not surprising that they began to have trouble working with foreigners. As funds became tighter there were fewer opportunities. With fewer opportunities, the call from scientists soon arose that U.S. ventures should be more restricted to U.S. scientists. Concomitantly, as less technology was being developed in the United States, came the call (from the military and Congress) that the United States should restrict foreign access to all advanced technology ventures.

NASA began looking to international participation during the post-Apollo program as a way to strengthen its own precarious domestic financial position. For the first time, NASA moved from the primarily scientific and technological motivations reflected in the guidelines for space science cooperation—which called for each cooperative venture to stand on its own scientific merits—to a more politically oriented approach. Political considerations moved from being a program or project benefit, to a motivation. Internationalizing a space project was then for many years assumed to lend stability and a higher degree of assurance of continuity than available to solely national projects. Examination of the first large-scale NASA-ESA cooperative project, Spacelab, points out some of the problems encountered under this new approach (Chapter 5).

Also during the second period, space became a tool of detente. The Apollo-Soyuz Test Project (ASTP) afforded cold war adversaries the opportunity to work together toward a highly symbolic "handshake in space" (Chapter 6). The pro's and con's of the project are debatable from both technical and political perspectives. What we can learn from it, however, is perhaps more important in terms of what *not* to expect from a cooperative space venture. Because the Soviets did not change their fundamental foreign policy as a result of ASTP, as some officials in the United States believed they should have been expected to, is not a reason to call the project a failure. Expecting cooperative space projects to impact foreign policy on a broader scale in the short term is setting oneself up for disappointment.

The utility of ASTP was also debated in the context of NASA priorities. Could the money and equipment committed to ASTP have been better used, as defined in scientific terms, on some other project? Was the trade-off between political benefits and scientific benefits enough to warrant this clearly political project? Linkage politics had always been a factor for the United States in space relations with our friends, and now it seemed to become one for adversaries as well.

The premise that internationalizing a space project gave that project stability and assured continuance began to be questioned by other nations after the cancellation of the U.S. spacecraft for the joint NASA/ESA International Solar Polar Mission (ISPM). That unfortunate incident (Chapter 6) was an exception to an otherwise steady norm and clearly, international participation was still seen as politically beneficial in both receiving and maintaining Congressional and administrative support for a program. But ISPM gave ESA and other potential NASA partners a hard look at the reality of the U.S. budget process as it affects the space program. They learned lessons they would not soon forget.

Finally, with the advent of the Shuttle came a new emphasis on "space for profit." Prior to the Shuttle, NASA had enjoyed the luxury of pursuing science and developing spacecraft on a research and development basis. With the Shuttle, however, the U.S. government began to demand a return on its investment. The ramifications of this decision on NASA as an organization have been staggering, and sometimes very costly in terms other than monetary. "How to make money in space" became a question of rampant speculation, with estimates for potential profit being made based more on wishful thinking than practical business considerations (Chapter 8). Again, this promoted unrealistic and unhealthy expectations about the benefits of space ventures.

Placing the end of this second period at 1984 is somewhat arbitrary. It could just as easily and correctly have been placed at 1986, with the *Challenger* accident, as that event in some ways resulted in a fundamental change of attitude in the U.S. space program. Further, when the Shuttle resumed flying, it was in a very different environment than previously, in terms of the number of actors involved in space activity and the roles that they were playing. The decision was made to end this second period at 1984 because that was the year Reagan "got religion" regarding space; that was the year Phase A space station negotiations began; and by 1986, the Soviet Union had clearly had a change of attitude about the future of its space relations which would profoundly affect the United States and other countries.

The second period of space history was one of rapid and somewhat dramatic change in space relations. Hopefully, individuals and countries learn from experience over time. This was a time to learn from.

Chapter 5

The First Large-Scale NASA-ESA Cooperative Project: Spacelab

Defining Success

Douglas Lord, who served as the Spacelab program director at NASA during the development of Spacelab, says in the final chapter of his 1987 book entitled *Spacelab: An International Success Story*:

> It would be hard to argue the conclusion that spacelab development was accomplished successfully. . . .
>
> No program is ever 100% successful, but if any international space effort is to be called a success, Spacelab qualifies. As a result, Europe now has a manned space system capability and the U.S. has an excellent manned laboratory system to use with its Space Shuttle.[1]

Lord, a strong advocate of international cooperation, is astute in his analysis of Spacelab's success. The technical achievements of Spacelab as the end-product of years of dedicated work are indeed commendable accomplishments.

Yet Lord also asks some questions about which he says "I do not pretend to know the answers to these questions, nor do I know anybody who does."[2] The questions he posits are unquestionably difficult:

> Why are former participants in the Spacelab program such difficult negotiators today? Is NASA unable to accept the concept of ESA as an equal partner in the next venture? Were both NASA and ESA so overbearing and inconsiderate in their demands of each other during the course of the Spacelab program as to create a lack of trust for future joint programs? Why do so many people consider the Spacelab program as a negative example, as the way "not to do it."

There is a relatively clear distinction between the bases on which Lord evaluates Spacelab as a success—technically and as an end product—and those on which his unanswered questions focus—the means used to reach the end. Without presuming to be able to identify comprehensive or explicit answers to Lord's questions, an examination of the political aspects of the Spacelab program serves well to at least begin to understand how the program's "success" could be challenged.

Problems in Europe

By 1970 it had become clear that the two European space organizations that had been formed in 1964, ELDO and ESRO, needed attention. ESRO had achieved a certain amount of success with the development of its first scientific satellites: ESRO-I (polar ionosphere and auroral phenomena), ESRO-II (solar and galactic cosmic rays and particles), and Heos I (interplanetary magnetic field and solar particles).[3] All three of the ESRO spacecraft were launched by U.S. rockets in 1968.

But ELDO was clearly in trouble. As discussed in Chapter 2, the Europa rocket, an effort intended to give Europe launch independence from the United States, was far behind schedule and over budget. Recognizing the problems with their two entity space organizations, ELDO and ESRO, and the imperative to hang together or hang separately, ministerial level meetings began in July 1970 to reorganize and set European priorities for space activities.

By 1972, the United States had agreed to European participation in the post-Apollo program through Spacelab. The Federal Republic of Germany wanted to accept the American offer of cooperative efforts, as it fit in well with their particular interest in manned flight programs. France, however, wanted priority placed on the development of an independent European launcher. To accommodate the differing interests of the ELDO/ESRO members, a compromise was reached whereby a "package deal" was accepted at the European Space Conference in 1973, which also instituted the creation of the European Space Agency (ESA). The package deal included priority being given to both Spacelab and Ariane—Spacelab to accommodate the German desire for participation in a manned space program and Ariane, a heavy launcher, to satisfy French interests in an independent space program.

Dependence on U.S. launch vehicles had also become a particularly acute issue after the first European (Franco-German) communications satellite, Symphonie, ran into problems with the U.S. government. As an operational communications satellite, Symphonie was viewed as competition to INTELSAT, so the United States mandated that Symphonie could be used for experimental purposes only. There were actually two satellites involved, Symphonie A and B, launched in 1974 and 1975, respectively. The French were particularly upset by what they perceived as the United States being able to in some ways dictate the direction of future European programs through launch restrictions. This increased their desire for autonomous European launch capabilities.

The European Space Agency

On 30 May 1975 the Convention for the Establishment of a European Space Agency was signed by Belgium, Denmark, France, Germany, Italy, the Netherlands, Spain, Sweden, Switzerland, and Britain, all former ELDO/ESRO members. Australia, which was a member of ELDO, decided it did not wish to sign the Convention while Ireland, a member of neither organization, expressed a desire to join and did so as an original member. In addition, two countries, Austria and Norway, became associate members in 1979 and 1981, respectively, and full members as of January 1987. Finland became an associate member in 1987 also. Canada has had a special cooperation agreement with ESA since 1981. ESA is headquartered in Paris with a staff, including personnel at various research and operation centers, totaling approximately 1,900.

The policy-making body of ESA is the Council, consisting of official delegates from the Member States, which meets about five times annually. The Council also meets at the ministerial level, to replace the meetings of the European Space Conference. The ministerial level meetings are held far less frequently, usually on matters of setting long-term policy.

Within ESA, there are two types of programs, mandatory and optional. Included in the mandatory programs are the scientific programs and contributions to the general budget, at a level proportionate to GNP shares. All other programs, specifically the applications program (e.g., Ariane and Spacelab) are optional. Each Member State has one vote in the Council, except when the discussion focuses on an optional program, and then only those who have chosen to participate vote. ESA resources for the mandatory activities are decided for a five year term, by unanimous vote of the Council. All other votes require a two-thirds majority.

ESA projects are funded differently than projects in the United States. Basically, once an ESA project is approved, it is funded for its lifetime. If, however, costs go above a predetermined percentage of the estimated total cost (usually about 20 percent), then approval for the expanded project must be obtained. For example,

Spacelab cost ESA about 40 percent more than the originally approved cost. Hence Member State approval had to be obtained to continue the project. In the case of Spacelab, Britain decided it could not continue participation at its previous level, and Italy picked up the additional British share of the project. So there are a few circumstances which could lead to the cancellation of an ESA project after approval for funding, but that has never occurred. The system used by ESA encourages accurate original cost estimates, and gives ESA projects the luxury of an almost guaranteed resource base after project approval.

Generally speaking, ESA operates on the basis of "juste retour." That is, the distribution of ESA contracts is determined according to the level of financial investment provided by the Member States. Initially, this was defined as not for individual contracts, but on a cumulative basis. Over the last several years Member States have insisted, however, that distribution be dealt with on a program by program basis. Clearly there are problems with this system as there are with any other. It is difficult to assure the equitable "industrial return" to a country, while still paying attention to the competitiveness of the industries involved and the quality of work. Specifically, the larger countries often have a certain competitive advantage due to their developed industrial base, giving rise to situations where special measures become necessary to increase the industrial return for some of the small countries.

In 1976–77 when Ariane was emerging from the R&D phase and was ready to enter commercialization, some Member States, particularly Germany, argued that an international public body like ESA was inappropriate to operate or market a launch vehicle. The structure of the organization would make decision making too complicated and bureaucratic. Further, the underlying reason for concern was that the ministries from the Member States were willing to fund R&D, but the money to move into production was more than they were willing to commit. Therefore, ESA began to work on the principle of "no profit, no loss," where ESA takes on research only, and as projects approach commercialization, they are turned over to industrial concerns. Such has been the case with the development of the Ariane launcher by ESA. Once declared operational, manufacturing, marketing, and operations were turned over to the private company—although one created by the French government—Arianespace.

Spacelab

Since the Apollo program, the domestic plight of NASA may be summarized as a constant battle with tight budgets and lack of consistent political and public support. Therefore, bringing an international dimension into the U.S. space program has been a necessary and desirable means of gaining and sustaining support within both the executive branch and Congress. It was quite a successful strategy until the Reagan administration.

The motivation behind ESA participation in Spacelab was as a means to an end regarding their desire for involvement in the NASA post-Apollo program generally—so as not to be left behind in the drive for technology. The United States encouraged Europe's Spacelab plans for reasons of political goodwill. Using these goals as the bases for evaluations, the picture becomes somewhat cloudy. Yes, ESA was able to become involved in the post-Apollo program (PAP), but whether or not NASA or the United States generated much political goodwill along the way is dubious.

In the days when NASA was making plans regarding "where to next," even before the Apollo program had ended, funding was a key consideration. Although overall space expenditures were actually on the rise, funds were increasingly going to the military side of the space program. Funds for civilian space ventures, as represented by the NASA budget, had started to decline in 1966. Public support, and hence Congressional budget support, for the space program began to flounder. "What are we spending the money for? We have reached the Moon so now let's solve our problems at home," was the domestic attitude that prevailed.

At the end of Project Apollo, the U.S. space program was typically left to the mercy of presidential interest. According to analyst James Everett Katz:

> Space policy . . . appears to be an area in which the president has unique power over its policy direction in the United States. Unlike many areas where the president is forced to accommodate and consult with numerous interest groups, decisions about space policy can be initiated spontaneously within the Executive Office of the president with little reference to outside concerns."[4]

NASA budgets did not do particularly well under either the Nixon, Ford, or Carter administrations. Although the Shuttle program was approved during the Nixon years, it was a reluctant approval accompanied by a strict warning to stay within the budget.

NASA planners had originally envisioned a reusable space vehicle and space station together as the logical next step in the U.S. space program. Funds, however, were insufficient for both. Subsequently, the space station idea was shelved, and it was decided that the Shuttle would go it alone as the next major NASA project. Even then, the Shuttle, if it had been kept to the original figures, would have had to have been built on a shoestring budget. NASA personnel cut corners every way possible to keep costs down, as they feared losing what support they had left if Congress thought them

going over budget. It was primarily because of military application interest that enough money was infused into the project from 1979 through 1981 to literally get it off the ground. Project cost was therefore a key concern of NASA. But to extrapolate that NASA went looking for international participation from Europe to keep costs down is a mistake.

Nixon and the Office of Management and Budget (OMB) rejected the Shuttle design originally proposed by NASA because of its $10 billion price tag. NASA went back and redesigned a $6 billion version. With the substantially lower price tag, and estimates of revenue to be derived from Shuttle payloads, NASA Administrator Dr. James Fletcher persuaded the Nixon administration to buy a fleet of four Orbiters.

NASA initiated talks and urged ESRO participation in the PAP for a very simple reason—it looked good politically. "International participation" might give the whole project more credibility and hence durability. Use of "international cooperation" as a public relations maneuver was not new to the U.S. space program, dating back to early slogans of "space for all mankind" and "space for peace" used to differentiate the civilian-run U.S. space program from the Soviet space program under the control of their military.

When NASA approached ESRO regarding participation in the PAP, ESRO had hoped for a partnership role, if only a junior partnership. ESRO wanted to provide some integral part of the Shuttle. Until then, Europeans had been satisfied to define cooperation as meaning to fly scientific experiments on U.S. spacecraft. Now, however, they wanted to move beyond the scientific aspects of cooperation and into the technological. Their dismal track record with the Europa left them in a poor bargaining position though, with little technical credibility.

NASA clearly was not interested in having the Europeans build *part of the Shuttle*. That seemed too technically risky. European participation was not intended to save money for NASA by having the Europeans contribute to the Shuttle itself. Rather NASA intended to let the Europeans participate in the PAP generally, to enhance the capabilities of the Shuttle system through a nonessential piece of hardware and to strengthen the program generally by being able to call it "international."

ESA first offered to build a reusable orbital transfer vehicle, called a space tug, as its contribution to the PAP. Objections from the U.S. defense sector, however, over possible technology transfers, halted that idea. It is interesting to note that a much-needed orbital transfer vehicle still has not been built. As an alternative, a scientific laboratory was suggested as the European contribution to the PAP. Spacelab had been conceived as part of the Shuttle–space station working unit plan, and when the space station plans were shelved, Spacelab was to go with it. The Europeans agreed to build it as their foot-in-the-door to the program.

The Spacelab (SL) agreement between ESRO-NASA was made official by a Memorandum of Understanding (MOU) dated August 14, 1973. The MOU put heavy emphasis on "non-completion" contingencies, as many NASA people did not believe that ESRO could actually follow through Article IX reads,

> ESRO will turn over to NASA without charge and without delay all drawings, hardware and documentation relating to the SL if ESRO abandons the development of the SL for any reasons, or ESRO is otherwise unable to deliver the SL flight unit prior to the first operational shuttle flight, or the completed SL does not meet agreed specifications and development schedules.

Although this contingency did not occur, ESRO/ESA virtually handed over everything in the end anyway. A European consortium of firms was designated responsibility for developing Spacelab in 1975 by ESRO. The consortium was led by the West German firm VFW-Fokker/ERNO (56 percent), joined by Aeritalia, Sener, Fokker, Matra, and British Aerospace.

Also per the MOU, NASA got full control of the SL unit (which actually consists of two sets of flight hardware, one engineering model, and the accompanying ground support equipment) after delivery. ESRO/ESA got one-half of the first Spacelab flight free (a joint flight with NASA), and after that they have had to pay to use it like any other customer. ESA-NASA cooperation on Spacelab was for all intents and purposes limited to the developmental phase. See Figures 5.1 and 5.2.

There are two general schools of thought in Europe about the Spacelab agreement. Some people feel that ESA got a bad deal from NASA; others feel that ESA basically negotiated a bad deal for itself and accepted it as the price to pay for entering the manned space program. Not all Europeans were happy with this agreement when it was made in 1973. According to French analyst Albert Ducrocq:

> That agreement was disastrous for the Europeans. We then denounced it vigourously, particularly during a memorable, and rather dramatic, debate at the FNAC (National Federation of Consumer Associations) where we stressed the 2 dangers resulting from the position adopted by the Europeans. We remember having been that day alone against all the others.[5]

The majority of Europeans though, especially the Germans, felt that the price was worth the technological experience to be gained. This is not to say, however, that they intended Spacelab to establish a precedent concerning working relationships between NASA-ESA

for the future. Indeed, Spacelab was later referred to by even its staunchest European supporters as the largest gift to the United States since the Statue of Liberty—and a situation not likely to be repeated soon.

Success or Failure?

Although one view is that NASA took advantage of Europe's poor bargaining position during Spacelab negotiations, NASA actions could also be looked at as "prudent." Working on a limited Shuttle budget, with questionable domestic support, NASA could not afford to risk the provision of an essential component to an operation with a background of failure. This reinforces the idea that political goodwill was its reason for cooperation. NASA did little to hide its attitude toward ESRO/ESA either. When NASA people went to the ERNO plant in Bremen, West Germany, in the early days of Spacelab work, many were quite condescending. In the opinion of at least some of the German engineers there at the time, the behavior of certain NASA personnel was much as though they had gone to an underdeveloped country.

As anxious as ESRO/ESA was to participate in the PAP, $370 million (1974 estimated cost of Spacelab) would certainly be a high price to pay to do so. In fact, the cost of Spacelab doubled from first estimates to final figures.[6] Included in the MOU was a provision for NASA to purchase at least one additional SL, beyond the unit to be turned over free of charge. It was also expected, based on estimated Shuttle flights and Spacelab usage figures, that several more SL units would be needed, to be procured from ESA at substantial profit. NASA did buy the required second Spacelab flight unit; the value accumulated over the years 1977–1984 to approximately $200 million including modifications and spare parts. The contract, the largest NASA ever made with a foreign entity, was signed with ESA, who then administered such with MBB/ERNO as again the prime contractor. It is not, however, expected that any further units will be needed or ordered.[7]

Except for ESA getting no further orders for SL units, making their financial benefit far less than they had hoped, both NASA and ESA did get much of what they wanted from the Spacelab cooperation, in terms of achieving their goals. The U.S. Shuttle program was carried through with far more domestic support than ever expected, and the Europeans gained a reputation as a worthy and capable partner in space ventures. What may be the more important consideration, however, is how ESA-NASA interaction on the Spacelab project, indeed the Shuttle program generally, will affect their mutual desires to work together further—on, for example, the Space Station.

Figure 5.1 Spacelab—a multidisciplinary platform for manned and unmanned missions. *Photo courtesy of European Space Agency.*

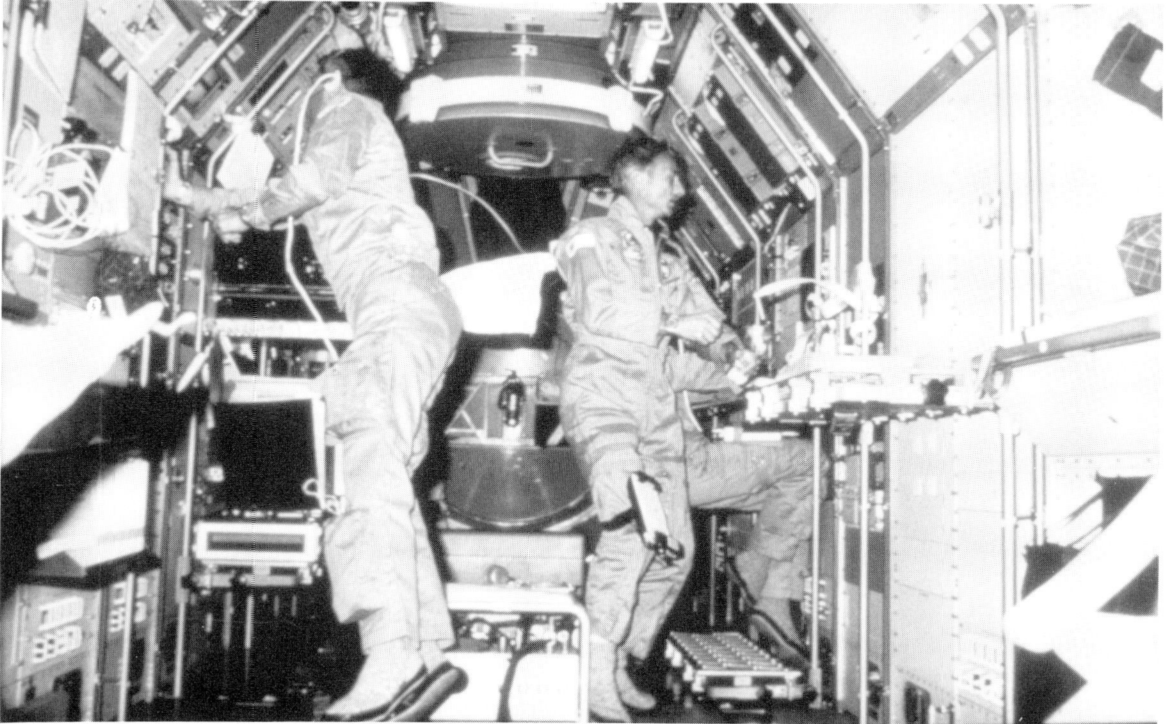

Figure 5.2 Working in Spacelab. *Photo courtesy of European Space Agency.*

Returning to the questions posed by Douglas Lord: "Why are former Spacelab program participants such difficult negotiators today? Is NASA unable to accept ESA as an equal partner? Were NASA and ESA overbearing and inconsiderate in their demands on each other? Why is Spacelab sometimes considered a negative example of cooperative efforts?" International program participants such as ESA are no longer naive in their expectations about NASA wanting to include them as an equal partner, nor are they still at a disadvantage in their negotiating position because of their lack of experience. They clearly want a return on their investment, in terms of scientific and economic benefits. NASA just as clearly expects to maintain the dominant role it has played in past cooperative ventures, and feels justified in that expectation because the U.S. financial contribution to joint projects is usually the largest, and because of the traditional U.S. technical expertise. Space is becoming subject to the same pressures of protecting national interests that are found in most political arenas.

Conclusion

Spacelab is sometimes cited as a negative example because it, like most learning experiences, was sometimes painful for the participants. Yet change, the kind of change that has been experienced in cooperative space project negotiations, is inevitable as part of the growth and maturing process. It can be a positive experience as long as players learn from it.

Whether or not NASA and the United States generally have learned from the past is, unfortunately, questionable. ESA, on the other hand, and other international partners on space projects must acknowledge that their "partnerships" with NASA are most often not a 50-50 split of expenses, and therefore will not result in an equal split of returns. The same premise that you-get-back-a-share-of-returns-commensurate-with-what-you-put-in that operates relatively effectively within ESA, ought to be extended to international projects as well. Another factor to be considered with that premise, however, is that NASA does not *offer* 50-50 project splits, so ESA—or any other partner—could not do 50 percent of a project if it wanted to. Therefore, if NASA wants international partners, it must make the deal attractive and worthwhile for those who might be interested in joining: cooperation implies all parties should benefit.

Endnotes

1. Douglas R. Lord, *Spacelab: An International Success Story*, Scientific and Technical Information Division, NASA, 1987, p. 401.
2. Ibid., p. 400.
3. *Europe's Future in Space*, The Royal Institute of International Affairs (London: Routledge & Kegan Paul, 1988), p. 73.
4. James Everett Katz, "New Directions Needed in U.S. Space Policy," in *International Space Policy*, Papp and McIntyre, eds. (New York: Quorum Books, 1987), p. 59.
5. Albert Ducrocq, "Cooperation ESA-NASA," *Air et Cosmos*, 21 January 1984, p. 37.
6. The price increase was caused primarily by the necessity of a parallel development with the Shuttle, i.e., Shuttle modifications had a great impact on the Spacelab. Also, ESA introduced modifications of their own to improve Spacelab capabilities. Cost overruns accounted for only about 30 percent of the price increase.
7. The Italian company Aeritalia has considered trying to initiate private commercial sales of Spacelab modules for customer use on the Shuttle. See: Jeffrey Lenorovitz, "Aeritalia Seeks Commercial Sales of Spacelab, Columbus Modules," *Aviation Week & Space Technology*, 1 July 1985, pp. 51, 53.

Chapter 6

Apollo-Soyuz: A Handshake in Space

The Cold War Thaws

The 1975 Apollo-Soyuz Test Project (ASTP) involving the United States and the Soviet Union has been the most visible, and for many people the most memorable, example of international cooperation in space. Although the United States has had much more extensive and continued cooperation with other nations, most notably European nations, ASTP was the most visually dramatic (handshakes in outer space on live television) and hence made the strongest and most prolonged impression.

For the ASTP, the United States launched a three-man crew in an Apollo spacecraft to dock with a two-man crew in a Soviet Soyuz. The spacecrafts docked on July 17, 1975, in orbit 120 miles above the Earth,[1] and for two days astronauts Tom Stafford, Deke Slayton, and Vance Brand exchanged pleasantries and gifts, and conducted several scientific experiments with cosmonauts Alexei Leonov and Valeri Kubasov. The clearly remarkable aspect of the mission was that it involved cooperation between two traditional rivals—in space and on Earth.

The case of the Apollo-Soyuz Test Project is one which clearly illustrates the influences of attitude, potential benefit, and political environment as factors on decision making regarding international cooperation in space. Although the details of the project itself are already well-known,[2] it is beneficial to reexamine that project from the position of those factors.

Political Environment

Although entirely interdependent, political environment, as the stage setting, is the factor that must be examined first. The era of detente, beginning in the late 1960s, created opportunities for cooperation between the United States and the Soviet Union impossible during the cold war. The signing of the Non-Proliferation Treaty in 1968 and the beginning of the Strategic Arms Limitation Talks (SALT I) in 1969 signalled a relaxation of tensions between the two super-

powers which would extend to fields other than arms control:

The Soviets stated that:

The atmosphere of the "Cold War" of the 1950s to 1960s precluded giving U.S.-Soviet cooperation in space the character of a constantly expanding process. It is not accidental that cooperative activity in the 1960s was limited to an exchange of information, contact between scientists, and individual experiments. . . .

Changes in the character of Soviet-American relations, and positive results of discussions on the highest level in Moscow and Washington, *allowed* (emphasis added) for a significant expansion of U.S.-Soviet cooperation in the research on and use of space.[3]

This indicates that were it not for the change in the political environment, cooperation would have been considered impossible from the Soviet perspective.

Considerations of prestige and competition would previously also have held cooperation on the scale of Apollo-Soyuz to be prohibited by the U.S. side. The ASTP mission was approved by the Nixon administration. It is interesting to note, however, that approval was not viewed as a show of support for the space program—one badly needed at the time—but as a show of support for Henry Kissinger's policy of detente. In fact, in January 1971 Kissinger met with the NASA team going to Moscow to negotiate a cooperative project, led by Acting NASA Administrator George Low. At that meeting Low requested clarification on the administration's position concerning an actual test mission using Apollo and Soyuz spacecraft. Kissinger replied that as far as the White House was concerned, Low had the authority to negotiate with a completely free hand in any area within NASA's overall responsibility.[4]

The political timing of the mission became increasingly important as plans progressed. At the later stages of the ASTP negotiations, it became clear that both sides were working hard to avoid or settle all differences and conflicts, and hence to hasten agreement. The reason behind the intense activity was the scheduled Moscow Summit in May 1972 between President Nixon and Soviet leaders. It seems that space was the first area in which successful negotiations between the United States and the U.S.S.R. had proved possible after the policy shifts inaugurated by President Nixon. For a time, it also appeared as though it might be the only such area. Therefore, the administration was anxious to have as much concrete evidence as possible to unveil at that Summit, showing that progress had actually been made toward better relations.[5] Clearly, Apollo-Soyuz not only required a change in political environment as a prerequisite for cooperation, but indeed was intended as a symbolic representation of the new political environment.

Attitude

Closely linked to the idea of political environment is that of attitude. Whereas political environment encompasses whether or not conditions are appropriate for cooperative ventures, consideration of attitude then focuses on whether or not either or both parties choose to work on a cooperative basis. An evaluation of attitude to a large extent involves an evaluation of potential benefits. Before that kind of secondary evaluation can be made, however, there has to be a prerequisite attitude that potential benefits could outweigh possible detriments. That attitude was not present during the cold war period. Although the rhetoric was sometimes otherwise, neither the United States nor the Soviet Union *politicians* really wanted to work together, which in turn limited the amount of cooperation the *scientists* were able to engage in. When the attitudes of the politicians changed in the era of detente, though, negotiations were rather quickly entered into to find an appropriate area for a cooperative venture.

Figure 6.1   Apollo-Soyuz docking. *Photo courtesy of NASA.*

Yet it is also possible to have an environment acceptable for at least limited cooperation, but have a particular U.S. administration or Soviet leadership decide for one reason or another that cooperation is not desirable. It is important to note that the attitude change which allowed ASTP to proceed was, on both sides, one of a very limited range. In fact, ASTP in retrospect was more or less a single, though dramatic, aberration to an otherwise unspectacular record of U.S.-Soviet cooperation in space.

Potential Benefits

The ASTP project was the result of evaluations of benefit on both a technical and political nature, by both the United States and the Soviet Union. The technical area which provided the raison d'être for ASTP was rendezvous and docking for space rescue. This area focused on a field which was at least theoretically nonthreatening to either side, and of interest to both. Originally, the mission called for a U.S. docking to a Soviet Salyut space station. When the Soviets insisted that the plans be changed, most observers assumed the Soviets' traditional desire for secrecy, especially concerning a new project such as Salyut, to be the reason. In retrospect, it appears that they were only trying to conceal schedule slips in the Salyut project: a two port Salyut, needed for the link-up, would not be available until two years past the scheduled date of 1975.[6]

The Soviet Union had some docking successes with their Soyuz flights, yet they had some problems also. In 1968, for example, an orbiting, unmanned Soyuz 2 was approached by a manned Soyuz 3, with the apparent intent of docking, yet no docking occurred. Later, in October 1969, a tandem flight of three manned spacecraft occurred and two were expected to dock. This never took place, which observers presumed to indicate problems.[7] This has led some analysts to posit that the Soviet acceptance of the ASTP, which was clearly out of line with their past policy, was prompted by the need to improve their docking technology. Their political and technological reasons, however, were not necessarily mutually exclusive. It is far more likely that the Soviets saw the opportunity to enhance their technical knowledge in an area where they were having problems, while gaining significant political benefits as well. Furthermore, according to at least some analysts, the technical benefits that were to be gained by the Soviets were not limited to the area of docking.

The political benefits to be reaped by ASTP for the Soviet Union included the opportunity for the Soviet Union to present itself as a technological equal of the United States. Negotiations for ASTP were going on at the same time as the United States was enjoying the success of the Apollo missions. It did not escape the Soviets that working with the United States on such a project might numb some of the public sting of "losing" the Moon race, although the Soviets continually claimed they were never in it.

Discussions for development of a common docking mechanism began in 1970. They were part of a larger agenda of potential cooperative space activities that the United States had been pursuing, with varying degrees of intensity and commitment. In January 1971 an agreement between NASA and the Soviet Academy of Sciences was reached, involving coordination of space activities, data exchanges, and a lunar sample exchange. This agreement served as the foundation for the 1972 intergovernmental *Agreement Concerning Cooperation in the Exploration and Use of Outer Space for Peaceful Purposes*, signed by Soviet Premier Alexei Kosygin and U.S. President Richard Nixon at the 1972 Moscow Summit.

The Moscow Summit was the first time an American president had officially visited the Soviet capital, evidencing the change in political environment. Subsequently, eleven bilateral agreements for scientific and technical cooperation were to result from the Summit discussions, including the agreement for cooperation in space. Article 3 of the agreement specifically dealt with ASTP.

> The parties have agreed to carry out projects for developing compatible rendezvous and docking systems of United States and Soviet manned spacecraft and stations in order to enhance the safety of manned flights in space and to provide the opportunity for conducting joint scientific experiments in the future. It is planned that the first experimental flight to test these systems be conducted during 1975, envisaging the docking of a United States Apollo-type spacecraft and a Soviet Soyuz-type spacecraft with visits of Astronauts in each other's spacecraft. The implementation of these projects will be carried out on the basis of principles and procedures which will be developed in accordance with the Summary of Results of the Meeting Between Representatives of the US National Aeronautics and Space Administration and the USSR Academy of Sciences on the Question of Developing Compatible Systems for Rendezvous and Docking of Manned Spacecraft and Space Stations of the U.S.A. and the U.S.S.R. dated April 6, 1972.

The United States saw the benefits to be gained from such a project as primarily political, as evidenced from the fact that the docking system to be developed was not going to significantly benefit the United States since the ASTP was to be the last time that an Apollo-type spacecraft would be used. This makes the political benefit to be had from this venture the target of much evaluation.

From a critical perspective, ASTP has been referred to as a "wheat deal in the sky" and a "quarter-billion-dollar space handshake";[8] evidencing doubts about the utility of the mission beyond the political benefits. The technology transfer issue has also been the subject of debate. Supporters of ASTP assert that there was little significant technology transfer. In 1984 space analyst Marcia Smith said "it is difficult to point to a single example of new space technology being used by the Soviets that might have come from their experience with ASTP."[9] Whatever the Soviets may have learned about American space techniques was most likely already available to them anyway through the open literature, especially since the Apollo-Saturn technology being used was already relatively old."[10]

But for critics the key operational definition is the word "significant." The Soviets were certainly able to observe, and to at least some degree assimilate into their program, American procedures concerning mission planning and operations. If one considers remodeling the Soviet mission control center to resemble mission control at Johnson Space Center, as the Soviets did,[11] to be a significant technology giveaway, then significant technology giveaway did occur.

Supporters say the symbolic value of the ASTP was well worth the various "costs" incurred, and that the United States learned about previously unknown aspects of the Soviet space program, although it would not be until approximately 1984 that the Soviets would open their program to the West (see Chapter 10). There has been no pretense though that the United States came away from ASTP with any new technical expertise.

Even accepting the premise that there were political benefits to be gained from ASTP and that technology transfer was minimized, the issue still stands as to where the benefits gained outweighed the costs incurred, in dollar terms. Opponents argue that ASTP was a wasteful use of scarce U.S. space funds; that money could have been used on a project that would help the United States on a long-term learning curve. Critics also objected to the one-time nature of ASTP, with no follow-up missions. It is important to note, however, that by 1978 the political environment had changed to the degree that any type of additional joint manned missions would not have even been considered. Some NASA people were also upset that ASTP used the last Saturn 1B launcher—which also could have been used for another flight to Skylab, a project with considerable internal NASA support and one where there was a strong feeling that it was underutilized.

Supporters argue that it is difficult to quantify the value of good will. Part of the problem too stems from the fact that it is difficult to calculate the exact cost of ASTP for either the United States or the Soviet Union. According to information contained in a technical memorandum put out by the Office of Technology Assessment, ASTP cost a total of $214.2 million. However, it goes on to explain that:

The existing Apollo Command Module and Saturn 1B launch vehicle, valued at $100 million, were transferred to the project at no cost from the completed Apollo program. (Similar leftover Apollo hardware was donated to the National Air and Space Museum.) Substantial additional support costs may have been incurred by NASA for ASTP which did not show up as a direct charge to the project. Soviet planners did not publicize their ASTP budget, which precludes a dollar-to-ruble comparison. Soviet costs included flying a practice mission, Soyuz 16, which went through all the maneuvers required for docking, and committing a backup Soyuz.[12]

There are also the practical benefits of ASTP to be considered in any evaluation of the project. Analyst James Oberg pointed out that the Apollo flight went on for a week after the Russians left, during which time the Americans carried out extensive experiments. Further, the timing of the ASTP mission, between the last of the Skylab missions and the beginning of the Shuttle missions, allowed NASA:

> to retain a cadre of experienced space specialists through what otherwise would have been a serious space-flight activities gap—during which most of them would have been laid off (and subsequent rehirings and retrainings would have been very expensive) and most of the valuable experience would have been lost for good.[13]

Fundamentally though, weighing economic and technical costs and benefits becomes a moot exercise, since political symbolism was the primary U.S. motive behind the project. Therefore, success or failure ought to be judged in those terms. Did the United States achieve what it wanted to with ASTP, and were the political goals realistic? From there, we ought to be able to learn something for the future.

Conclusion

First and foremost, the lesson to be learned from ASTP is not to impose unrealistic expectations on cooperative space ventures, otherwise the project and those involved are being set up for disappointment and dooming the mission to "failure" in at least some sense. Remarks made by Charles E. Horner, deputy assistant secretary for scientific and technology affairs of the U.S. State Department, before the Senate Foreign Relations Committee in September 1984, suggest that at least some people may have tried to use such criteria to evaluate ASTP:

> The first observation I wish to make about the nature of our experiments to date in this area is that, in the Department of State's view, cooperative projects in space have not had any direct influence on Soviet geopolitical and military aims. For example, the Apollo-Soyuz project did nothing to restrain the Soviet Union's adventurism in Angola, or the invasion of Afghanistan or to prevent the

massive and unprecedented buildup in Soviet strategic arms which took place in the 1970s.[14]

Any expectation that cooperation between the United States and the Soviets on a space project of this type would lead to a fundamental change in Kremlin philosophy or activities would be ludicrous.

Cooperative programs should be designed and approved based on their scientific merits and evaluated on the same basis. That political benefits will accrue should be viewed as an added, and not inconsequential, value.

Endnotes

1. This low altitude presented an impressive view for the U.S. astronauts. Such low altitudes had been avoided before because of communications problems and high air drag, but it was as high as the Soviet craft could fly, so the United States had to go along with it.
2. See, for example: Dodd L. Harvey and Linda C. Ciccoritti, *U.S.-Soviet Cooperation in Space* (Miami: Center for Advanced International Studies, U. of Miami, 1974), Edward Ezell and Linda Ezell, *The Partnership: A History of the Apollo-Soyuz Test Project*, NASA SP-4209 (Washington, D.C.: NASA, 1978), and *U.S.-Soviet Cooperation in Space* (Washington, D.C.: U.S. Congress, Office of Technology Assessment, OTA-TM-STI-27, July 1985), pp. 23–32.
3. G. S. Khozin, *S.S.S.R.-S.Sh.A.: Orbity kosmicheskogo sotrudnichestva* [U.S.S.R.-U.S.A.: Orbits of Space Cooperation] (Moscow: Mezhdunarodnye Otnoshenia, 1976), p. 6. quoted in *U.S.-Soviet Cooperation in Space* (Washington, D.C.: U.S. Congress, Office of Technology Assessment, OTA-TM-STI-27, July 1985), p. 24.
4. Edward Ezell and Linda Ezell, *The Partnership: A History of the Apollo-Soyuz Test Project*, NASA, SP-4209 (Washington, D.C.: NASA, 1978), p. 126.
5. Dodd L. Harvey and Linda C. Ciccoritti, *U.S.-Soviet Cooperation in Space* (Miami: Center for Advanced International Studies, University of Miami, 1974), p. 228.
6. James E. Oberg, *Red Star in Orbit* (New York: Random House, 1981), p. 140.
7. *U.S.-Soviet Cooperation in Space* (Washington, D.C.: U.S. Congress, Office of Technology Assessment, OTA-TM-STI-27, July 1985), p. 23.
8. Oberg, *Red Star in Orbit*, p. 139.
9. Marcia Smith, "America's International Space Activities," *Society*, January/February 1984, p. 19.
10. Oberg, *Red Star in Orbit*, p. 143.
11. Smith, "America's International Space Activities," p. 19.
12. *U.S.-Soviet Cooperation in Space* (Washington, D.C.: U.S. Congress, Office of Technology Assessment, OTA-TM-STI-27, July 1985), p. 27, ft. 31.
13. Oberg, *Red Star in Orbit*, p. 144.
14. *East-West Cooperation in Outer Space*, hearings before the Senate Committee on Foreign Relations, September 13, 1984 (Washington, D.C.: U.S. GPO, 1984), p. 71.

Chapter 7

The International Solar Polar Mission

A Negative Precedent

As the saying goes, you learn from experience. Unfortunately, some of the experiences that Europe had with the United States in space cooperation during this second, or tutorial period, were not particularly heartening.[1] Probably the most important lesson that they learned was that the U.S. budget process makes it impossible for NASA to guarantee the continuation of an international project beyond a yearly basis. This lesson would not be forgotten by the European Space Agency (ESA), or other international players in the space community in later negotiations, particularly those concerning the space station. One of the most frequently cited negative precedents, in terms of cooperative projects that encountered problems, concerns the 1981 cancellation of the U.S. spacecraft from what had been planned as a two-spacecraft U.S.–ESA International Solar Polar Mission (ISPM), now known as Ulysses.

Origins of the ISPM Mission

The NASA Fiscal Year (FY) 1979 budget request submitted to Congress in January 1978 contained five "new start" projects. Included in these was the International Solar Polar Mission, a joint NASA–ESA mission designed to study the polar regions of the Sun in an effort to understand better the Sun's influence on the Earth's environment. As originally planned, the ISPM was to have two spacecraft, one supplied by ESA and the other by NASA. They were to take measurements and to make observations as they orbited outside the ecliptic plane (the plane of the Earth's orbit around the Sun) over the solar polar regions.

The two spacecraft were to be launched in February 1983 by the U.S. Space Shuttle. Using the inertial upper stage (IUS), the spacecraft would be launched from the Shuttle almost directly away from the Sun to intercept the orbit of Jupiter about 16 months after launch. The uniqueness of the mission lay in the concept that each spacecraft would be accelerated out of the ecliptic plane by utilizing Jupiter's exceptionally strong gravitational field. The two spacecraft were to separate at Jupiter, with one proceeding to an orbit over the north Solar pole while the second simultaneously orbited over the south Solar pole. Plans for the mission, however, were changed almost as soon as they were announced.

Problems in Congress

The Senate Appropriations Committee Report on the NASA FY 1979 budget provided $13 million for ISPM and projected a total ISPM cost between $190 million and $230 million. As launch was scheduled for 1983, that meant that between $177 million and $217 million had to be allocated to the project over the next four years. But in order for ISPM to be launched, the Shuttle had first to be operational, and that was clearly NASA's first priority. Therefore, in the final FY 1979 budget, a reserve fund for Shuttle design, development, test, and evaluation (DDTE) was included to cover potential shortfalls in 1979 Shuttle funding needs. This fund was created by reducing the Space Telescope budget by $15 million, the Jupiter Orbiter Probe (later

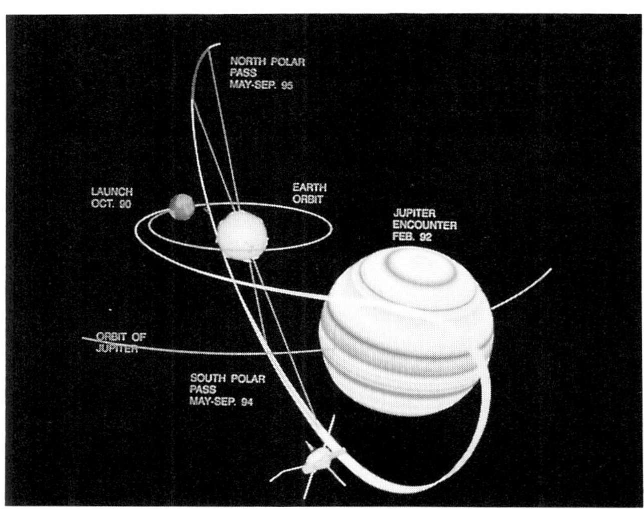

Figure 7.1 Ulysses flight path. *Photo courtesy of European Space Agency.*

Figure 7.2 Ulysses being prepared for launch. *Photo courtesy of European Space Agency.*

Galileo) budget by $10 million, and the ISPM budget by $5 million.

In December 1979, the chairman of the Senate Appropriations Subcommittee that had jurisdiction over NASA appropriations wrote to NASA Administrator Robert Frosch, suggesting that ISPM be delayed for two years. The reasons given for the proposed postponement were (1) to reflect the delays in Shuttle development and (2) because the committee was concerned that the IUS necessary to send the two spacecraft on the flight path would not be adequate for the job and that NASA should develop a high-energy upper stage (Centaur) instead.[2]

This stretchout request began a series of problems for ISPM which culminated with the cancellation of the U.S. spacecraft. A few months later the House Appropriations Committee recommended in the 1980 Supplemental Appropriations Bill that ISPM be terminated, and that some of the money already appropriated for FY 1980 be rescinded. This recommendation was made after review of the proposed FY 1981 budget in early 1980. For FY 1981, NASA initially requested a budget totalling $5.7 billion. The request was later amended as a result of the Carter administration's election-related desire to present a more balanced FY 1981 budget. The March 1980 revised budget request of $5.5 billion was $219 million below NASA's original request. Shuttle development funds were *not* reduced in the revised budget. At $1.9 billion, those funds were the major item in the NASA budget request. Protecting the Shuttle budget while decreasing the overall budget by over $200 million led to reducing space science programs by $107 million. Thus the total amount allocated to space science in FY 1981 was $40 million less than in FY 1980. Included in that $107 million reduction was $43 million cut from the ISPM; this necessitated a two-year launch delay. This cut from the ISPM budget and the consequent launch delay had been discussed and agreed upon by NASA and ESA officials.

The House Appropriations Committee, however, decided to cancel the program altogether. The committee issued the following explanation:

The committee has rescinded $15,000,000 of research and development funds and directs that the solar polar mission be cancelled. . . . Initial funding of this program was provided in the amount of $12,500,000 in 1979. The 1980 current estimate is $47,900,000.

The revised 1981 budget proposes that the solar polar mission be slipped two years—to 1985. This will have the effect of saving $43,000,000 in 1981—but delaying the mission for two years will add *at least* $150,000,000 to the total cost of the mission.

In short, the budget revisions are inefficient and simply delay projects which result in long-term cost growth.[3]

The European Space Agency reacted to the proposed project cancellation by mounting a strong diplomatic protest. But it was Representative Don Fuqua (Democrat, Florida), chairman of the House Science and Technology Committee, who was able to quash the proposed rescission by successfully arguing that the cancellation of the funds would constitute legislation in an appropriations bill—a violation of House rules.[4] Thus the $15 million rescission in FY 1980 for ISPM was deleted. However, the reprieve for ISPM was only temporary.

These cuts affected ISPM and the NASA budget almost incidentally. The budget problems experienced by NASA in FY 1979 and FY 1980 came basically from the need to provide supplemental shuttle development funds. In FY 1981, the Carter administration made overall budget cuts in an election year attempt to bring down the federal deficit.

Reagan Budget Philosophy

The whole budget process and attitude fundamentally changed with the November 1980 election of Ronald Reagan as president and Reagan's appointment of David Stockman as director of the Office of Management and Budget (OMB). Prior to 1980, the budget process tended to be a struggle between the parts and the whole, where the sum of the parts tended to drive the whole. That is, individual programs and other budget items were totalled to derive a final budget figure. Stockman's approach when he took over OMB in January 1981 was first to set the total budget figure and then to divide the total among the parts. Stockman introduced a system whereby the White House effectively began to put together and control budget initiatives, a power which had previously rested with the executive branch agencies and various subcommittees of Congress.[5]

When Jimmy Carter left office, he had left a financial blueprint for FY 1982 to hike taxes and to allow funding for most federal programs to continue at then current or even increased levels. Speaking to reporters on January 15, 1981, Stockman said that the president-elect planned to revise Carter's proposals "from top to bottom" and that Reagan would make "very, very major cuts."[6] Although the specifics were not known, the general expectations were clear.

The original FY 1982 budget submission for the NASA space science program, as part of the final Carter budget presented on January 16, 1981, was for $757.7 million. Of that figure, $58 million was for ISPM. The amended budget submission for space science after the Reagan administration took over was

for $584.2 million. For reasons discussed below, this cut forced NASA to cancel its ISPM spacecraft. The amended budget was presented by the president as part of the State of the Union Message on Thursday, February 19. In just over one month, drastic cuts had had to be made, and made very quickly. This had allowed NASA officials little time to carefully examine programs and options. Further, the discussions on the budget within the U.S. government generally, including those concerning NASA, were far more secretive than in other years. This strict secrecy policy made anything more than general speculation about the impending budget cuts nearly impossible for outsiders, even those who ordinarily had excellent contacts and access to information within government agencies like NASA.

ESA Reaction to ISPM Spacecraft Cancellation

Director-General (D-G) Erik Quistgaard of ESA was informed of NASA's decision to eliminate the U.S. ISPM spacecraft just a few hours prior to Reagan announcing his budget cuts. Quistgaard was reached by telephone in the early morning in Noordwijk, the Netherlands. He spoke with Acting NASA Administrator Alan Lovelace and Director of International Affairs Kenneth Pedersen about the spacecraft cancellation. He was told that the elimination was necessitated by a $500 million budget cut in the NASA budget overall. Quistgaard was also informed that the launch of the remaining European spacecraft would be postponed from 1985 until 1986.

On February 23, 1981 ESA D-G Quistgaard relayed to the Member State delegations the contents of a February 20 telex from Lovelace which elaborated on the reasons for NASA's decision.

As I indicated to you in our telephone conversation yesterday, the administration's budget for FY 1982 requires severe cuts in the full range of NASA's programs. Because work on vital shuttle activities must continue, we have been forced to cancel or otherwise forego a number of major programs in the science and applications areas. . . .

We have endeavored, and will continue to endeavor, to honor our international commitments to the fullest extent possible. Nonetheless, the deep budget cuts have necessitated cancellation of part of the joint NASA/ESA ISPM mission, namely the U.S. spacecraft which was to have participated in the solar polar mission.

In view of the scientific importance of the solar polar research, we hope that ESA will continue with the mission which can now be launched in 1986 on a shuttle/centaur and that we will be able to maintain its cooperative nature.

As I indicated to you yesterday, the NASA budget will permit support of the remaining spacecraft, including the U.S. experiments previously planned for the ESA spacecraft.

* * *

> I want to assure you that cancellation of the U.S. spacecraft in the ISPM mission is taken with great reluctance and was unavoidable given the broad and deep cuts occurring throughout NASA and throughout the U.S. government budget. . . .[7]

The Scientists' Lobby

ISPM was one of three projects focused on by the U.S. Jet Propulsion Laboratory (JPL) during Bruce Murray's tenure as director between 1976–1982: the Jupiter Orbiter Probe (JOP), later named Galileo, and a mission to Halley's Comet being the other two. All were planetary projects, an area of space science which suffered almost devastating funding cuts between fiscal year 1973 and fiscal year 1978. By fiscal year 1978 appropriations had dropped to less than one-fourth of their fiscal year 1973 level. This situation at times challenged the very survival of JPL and motivated scientists to become actively involved with lobbying for planetary exploration generally,[8] and in this case for the ISPM mission specifically.

According to J. David Bohlin, the program scientist for the ISPM/Ulysses mission at NASA Headquarters, the extent of involvement of the U.S. space science community in the fight to get the U.S. spacecraft reinstated should not be overlooked or underestimated. He says:

> Their letters and visits to Washington, and the recommendations to reinstate the U.S. spacecraft by NASA's science advisory committees at all levels, constituted one of the strongest lobbying efforts ever mounted by scientists with respect to a space program up to that point in time.[9]

Unfortunately ISPM was not the only space science mission which was challenged by budget cuts, which meant loyalties and the efforts of the scientists' lobby became divided. For JPL scientists, for example, both ISPM and the Halley's Comet mission were threatened, and both eventually cancelled.

After-the-Fact Negotiations

A meeting was held in New York on Monday, February 23, between NASA and ESA officials to discuss the U.S. spacecraft cancellation. Victorio Manno of the ESA Scientific Directorate and Wilfred Mellors, director of the ESA Washington office attended for ESA. Andrew Stofan, acting associate administrator for space science, and Kenneth Pedersen, director of international affairs, represented NASA. The NASA position at the meeting was that the cancellation was the result of severe budget cuts imposed on the Office of Space Science, but that NASA intended to continue to support the other elements of the cooperative mission related to the ESA spacecraft, specifically, the U.S. experiments, the launch, the radio isotope thermal generator, and the data retrieval and dissemination. The

ESA position was that the cancellation was a unilateral breach of the ISPM Memorandum of Understanding (MOU), that the cancellation was totally unacceptable and that ESA requested full restoration of the program to the level set out in the ESA–NASA MOU. The ESA officials also reminded NASA that when the ISPM project was accepted by the ESA Science Programme Committee in 1979, it was chosen in preference to a number of other, purely European missions because of the value ESA attached to transatlantic cooperation.[10]

At the February 23 meeting, NASA informed ESA that detailed federal budget proposals would not be sent to Congress until March 10. The ESA Management Board discussed the situation on February 24, and concluded that strong and immediate actions were necessary to achieve a full restoration of the two-spacecraft mission. Quistgaard thought that action at the political level by the individual ESA Member States had the best chance for success. He requested each Member State to make the strongest possible protest through its embassy in Washington— preferably at the ambassadorial level—against the decision taken by NASA. In view of the March 10 deadline, delegations were asked to give such actions the utmost priority.[11] An ESA press release on February 26 concerning ISPM said that "The Agency feels that if action is taken at top level, and within the next few days, there remains a good chance that ISPM will be fully restored within NASA." This optimism was unfortunately not well-founded.

In response to ESA's charge of abrogation of MOU responsibilities, NASA referenced Article 13 of the MOU which stated:

> It is understood that the ability of NASA and ESA to carry out their obligations under this Memorandum of Understanding is subject to their respective funding procedures.

Although technically Article 13 did give the United States an "out," *Aviation Week & Space Technology* quoted a European official as saying, ". . . but we did not think it would be used unless NASA was going bankrupt or something as serious as that."[12] Further, the point that the Europeans repeatedly stressed was that per Article 13, they had expected that NASA would at least try to get the funds, understanding that it was the prerogative of Congress to deny them. That NASA had not even included a request for ISPM funds in its budget request did not, in their view, live up to the terms of its MOU obligation, much less expected professional standards of conduct.

The ESA reaction to the U.S. spacecraft cancellation can best be understood as stemming from three sources: incredulity that an international space agreement would be cancelled at all; outrage because of the way the cancellation was done; adamancy in the demand for spacecraft reinstatement because several European and U.S. solar experiments would be grounded by the cancellation of the NASA spacecraft. That NASA acted with no prior consultation and with minimal warning to ESA officials concerned the Europeans as much as did the scientific and financial ramifications of the cancellation.

The incredulity regarding NASA's willingness to cancel an international program reflected ESA's stunned realization of a fundamental difference in attitude between the two organizations concerning the sanctity of international agreements. Within ESA, in a similar budget situation national space projects would have given way to ESA projects; ESA had assumed that the same was true for NASA—that international projects would be given precedence over national projects. ESA officials assumed that NASA would at least include ISPM in its budget request to OMB, and that OMB would pass it on to Congress for a decision on funding there.

From NASA's perspective, there were only three space science projects large enough to make the required cuts from: Galileo, a West German–NASA project; the Hubble Space Telescope, a multilateral project with strong support from the international scientific community; and ISPM. The OMB had first suggested that Galileo be cut, since it involved only one other country, rather than eleven. When NASA balked, wanting to protect Galileo (an orbiter-probe mission to Jupiter) because of its more central role in the planetary program, OMB washed its hands of the matter and insisted that NASA make the decision—one that clearly was going to result in ill feelings.

Attempts to Find a Compromise

On March 3, 1981 the three senior ESA Member State ambassadors, those from Italy, Switzerland, and Sweden, presented an *Aide Memoire* to the U.S. State Department on behalf of the eleven ESA Member States. It stressed the financial, scientific, and political consequences of the U.S. cancellation.

> . . . European investment in the overall programme amounts to approximately two hundred million dollars, equivalent to almost the whole of the Agency's annual budget for space science, have already been committed. This investment was judged against anticipated scientific objectives which would now be seriously degraded.
>
> Furthermore, the funds already expended by the European scientists participating in the U.S. spacecraft amount to approximately fifteen million dollars. This sum would be irretrievably lost.
>
> While the governments of the eleven Member States of the European Space Agency are aware of the desire of the U.S. government to contain public expenditure, they nonetheless believe that this must not be done at the expense of existing international agreements. They are furthermore convinced, especially in times of financial strin-

gency, that continued international cooperation is essential to carrying out cost-effective programmes with limited resources. The governments of the Member States therefore strongly urge that the U.S. government reconsider its position with a view to restoring full U.S. participation in the mission, and that this should not be achieved to the detriment of other international space projects.[13]

The NASA budget sent to Congress on March 10 included no provisions for the U.S. spacecraft portion of ISPM. In a telex sent March 17, 1981, to the Member States' delegations, Quistgaard said he did not consider the matter final and said that there was a "possibility, albeit small, that a solution . . . will restore the two spacecraft at the cost of a delay in the launch date."[14] The Europeans saw maintaining diplomatic pressure as the key to that possibility.

A compromise "solution" proposed by ESA was to have the United States provide a less expensive spin-stabilized spacecraft; this would eliminate the more complex despun platform of the originally proposed NASA probe. ESA said it might be willing to accept a one-year delay from the 1985 launch date with such a compromise, if ESA would incur no additional program costs other than those directly attributable to the European portion of the mission resulting from the delay in launch date.

ESA officials stated that the European aerospace industry could build for NASA a copy of the ISPM spacecraft which was already being prepared for ESA by Dornier. That could be done at a cost of not more than $40 million, as compared to the anticipated $100 million cost of the original U.S. spacecraft. European officials repeatedly stressed that they were not trying to push a European spacecraft, but merely trying to offer a feasible and less costly alternative.

Throughout March and April 1981, consultations continued among and between ESA, NASA, Congress, the State Department, OMB, and the White House Office of Science and Technology Policy (OSTP). On March 31, Quistgaard informed the Member State delegations that:

> I feel that the positive attitude indicated by NASA, together with the sympathetic understanding of our position expressed in the context of the Congressional hearings on NASA's budget, allows me to display some cautious optimism that we are nearing a satisfactory solution to restore a two spacecraft mission.[15]

On April 10, the State Department sent an *Aide Memoire* to the Washington embassies of the ESA members in reply to their March 3 demarche. The document offered strong rhetorical support, but made no promises. At this point a full spectrum of positive and negative signals were being sent between various U.S. and ESA officials; in the long run this very prob-

Figure 7.3   Galileo spacecraft. *Photo courtesy of NASA.*

ably gave ESA officials more encouragement than was actually warranted.

The general position taken by a number of U.S. actors in the negotiations can be characterized as sympathetic but noncommittal. Reporting back to the Member State delegations on the negotiations in an April 23 telex, Quistgaard said he had met with Congressman Edward Boland, chairman of NASA's Appropriations Subcommittee, and felt that Congress would view a restoration of a two-spacecraft mission sympathetically *if the executive branch were to propose it.* (emphasis added) Quistgaard's assessment of the executive branch position was that:

> a) any solution which would imply violation of the financial constraints imposed by the OMB for this and successive fiscal years was doomed,
> b) neither a solution based on a U.S. produced simpler spacecraft nor, a fortiori, that based on the original concept could be maintained within the above mentioned financial constraints,
> c) only a solution based on the purchase by the United States of a second unit of the spacecraft currently being developed by Dornier for ISPM could in all probability be envisaged within the limits.
> The OMB was, however, very wary even about this latter solution, because of a general preoccupation regarding the deficit of the federal budget in FY83 and also because little was known of the European conditions for a second Dornier spacecraft.[16]

Besides relaying his perceptions of the positions taken

by the various U.S. actors involved, Quistgaard also noted that ISPM was receiving a great deal of media attention, and attention by government officials in both Congress and the executive branch. That, he felt, would work in ESA's favor.

ESA officials felt that the option of a second Dornier spacecraft would be favorably viewed for several reasons. First, it could salvage most of the scientific content of the mission. Only one experiment, the coronograph, could not be accommodated on the simpler Dornier spacecraft. The option of using a Dornier spacecraft also would not, in all likelihood, violate OMB financial constraints and would still offer NASA considerable savings. Therefore, as Quistgaard put it, ESA was

> ... approaching a situation where the problem would cease to be scientific, technical or financial, and, instead, become purely political, on which grounds we feel, of course, more confident of a satisfactory outcome, due to the successful representations of our ambassadors.[17]

ESA's intention at this point was to alleviate OMB's hesitations about European conditions for a second Dornier spacecraft. The ESA proposition specifically was: (1) that the development would be under ESA management; (2) that the "go-ahead" would be given to Dornier by July 1981 at the latest; (3) that the experiments would have to be adapted to the spacecraft interfaces; (4) that the development would be compatible with a 1986 launch date; and (5) that the price to the United States would be fixed at $40 million, at 1981 prices, to be updated for inflation. The last point was considered essential in return for the first point, and to dispel U.S. fears over potential cost escalations later that would jeopardize the rigid budget constraints outlined.[18]

ESA felt that the Dornier option was a fair arrangement whereby ESA would get what it wanted, a two-spacecraft mission, and NASA could still meet its budget requirements. Also, willingness on the part of NASA to accept an alternative solution certainly would go far toward smoothing ruffled political feathers in Europe. For its part, ESA was willing to support the good faith effort of the United States by assuming the risks of potential cost overruns.

Attempts at Compromise Fail

On April 24, Acting NASA Administrator Alan Lovelace sent a letter to OMB Director David Stockman stating that the ESA proposal for NASA to buy a Dornier replacement spacecraft was scientifically acceptable, but that additional funds would be required in fiscal years 1983–86 to pursue this option. Lovelace asked for guidance on how to proceed. Between the time of Dr. Lovelace's request and Stockman's reply, discussions continued between ESA and U.S. government officials. From May 11–13, Erik Quistgaard met with policy-level officials at OMB, OSTP, the Department of State, and NASA, and with several members of Congress, Congressman Boland in particular. According to Quistgaard, the attitude of the State Department was seen as sympathetic, OMB was said to be taking a more "constructive attitude" though remaining noncommittal, and officials at NASA and OSTP expressed concern, but were reluctant to reach any decision before a new NASA administrator and deputy administrator took office. Prospects for an eventual solution were, however, seen by Quistgaard as "relatively good . . . due to the continuous pressure exercised and that it was essential that this pressure be maintained until a final positive conclusion was reached."[19]

Stockman finally replied to Lovelace in June that ESA representatives had been informed that "due to further likely cost growth on the space shuttle and other space projects which would exacerbate the already difficult fiscal year 1983 situation, no add-on was possible for ISPM." Stockman added that NASA had flexibility to reprogram funds from within existing budgetary resource levels.[20]

James Beggs took office as the new NASA administrator on June 27, 1981. The decision fell to him whether to rearrange NASA priorities to allow for a two-spacecraft ISPM mission. The "flexibility" referred to by Stockman meant cutting funds from another international program in order to pursue a two-spacecraft ISPM. In reality, therefore, NASA did not feel it really had much "flexibility" to rearrange priorities. The key consideration regarding ISPM was that without additional money and the associated support of OMB, NASA was unwilling even to try to proceed with the two-spacecraft mission.

NASA Administrator Beggs on September 4, 1981, informed Quistgaard that NASA would not include any request for funds for the second ISPM spacecraft in its FY 83 budget proposal which would be submitted to the White House in a few days. Quistgaard replied on September 9, reviewing the congressional and other support generated for the two-spacecraft mission. The congressional "support" that he was referring to was primarily rhetorical, although money had been added to the NASA FY 1982 budget by the appropriations conference for ten specified items, including ISPM. This was supposedly to allow NASA to maintain the two-spacecraft option. Quistgaard then went on to say that:

> the above elements [support], coupled with the fact that no in-depth consultation with ESA was held, as evidenced by the fact that we were never requested to expand on the European option, put me in a situation where I cannot even explain to our delegations the reasons for your actions.[21]

Quistgaard's frustration was obvious. He had sincerely felt that the second Dornier spacecraft option was fair to all parties and was being favorably considered in the United States.

Beggs finally gave the following reason for the NASA decision:

> While your offer for procurement of the second spacecraft was most constructive and greatly appreciated, my conclusion was that the elimination of the imaging system required by the approach substantially reduced the scope of the scientific results of a two-spacecraft mission. Thus, because of the science, my consideration of this issue focused on budgetary implications of a complete two-spacecraft mission with imaging. This simply proved to be unattainable in the current and foreseeable budget climate.[22]

Although Lovelace in April had said that a second Dornier spacecraft was scientifically acceptable, Beggs in September used scientific considerations as the official excuse for not supporting that option. Also, certain members of Congress and space committees questioned whether U.S. funds should be spent on a spacecraft built by European industry rather than what were considered "critical Congressional districts" in the United States. That was perhaps as much or more a consideration in evaluating the Dornier option as scientific acceptability.

Beggs did encourage ESA to continue with the mission without the U.S. spacecraft. He stated that NASA would

> use its best efforts to fulfill its remaining commitments in the original MOU. Thus, NASA intends to provide launch services, tracking and data acquisition support, and RTG and U.S. experiments for the European spacecraft, and other services as agreed in the MOU.[23]

The saga of ISPM was played out somewhat longer, through the summer of 1982, in relation to the submission of NASA's FY 1984 budget to OMB. The science community let it be known that they felt that some of the core science objectives intended to be accomplished by the cancelled ISPM experiments could be recovered by flying them on an in-ecliptic mission called the Solar Interplanetary Spacecraft (SIS). An ad hoc committee was established by the National Academy of Sciences (NAS) to examine that mission, in response to a congressional mandate which basically said that NASA could not unilaterally make radical changes in "approved" science missions without appropriate peer review. Clearly that premise did not work both ways, and NASA must have seen a certain irony in Congress' do-as-we-say-not-as-we-do attitude. The NAS did ultimately give the SIS mission a favorable endorsement, but at the last minute NASA Administrator Beggs decided not to include it in the agency's budget request.

After discussion of the scientific and financial ramifications of the U.S. cancellation, ESA officials early in 1982 announced their decision to proceed with the single ESA spacecraft. In July 1984, the mission was renamed Ulysses. The flight-ready European spacecraft is now scheduled for launch in October 1990, unless further postponements become necessary.

Problems and Policy

The handling and outcome of the International Solar Polar Mission was the exception to a general tradition of successful international cooperation on space projects. Although an aberration, it clearly illustrates one of the basic dilemmas for the United States of international cooperation on complex and expensive projects: that of the European (or any partner's) desire for iron-clad assurances from the United States regarding fulfillment of the U.S. portion of any project, versus what is possible within the realm of the highly political U.S. budget process.[24] The reality of the situation is that such guarantees are not possible. But it is also true that few international space programs have ever been cut once initiated.

The cancellation of the U.S. probe in ISPM was clearly the unfortunate result of an unprecedented set of circumstances in 1981. There was an acting administrator at NASA who was unwilling or unable to champion the project commitment, a continuing need

for supplementary money for the Shuttle which choked any flexibility in the overall NASA budget, and, most important, there was a new administration with a budget director and budget plan unlike anything Washington had experienced in a long time. Cuts had of course been made in the NASA and ISPM budgets during previous years, but they had been absorbed, rather than resulting in the cancellation of part of a signed international agreement. Further, the strictly enforced embargo order from the White House concerning the Reagan administration's changes in the FY 1982 budget made what would have been a bad situation under any circumstances worse by adding the element of surprise.

After ISPM, ESA officials began vocally to question whether MOU's were appropriate as legal instruments for international space projects. Indeed, after the cancellation of the U.S. spacecraft ESA officials repeatedly asked for legal "assurances" and "guarantees" that NASA would not back out from remaining ISPM obligations. Again, NASA was unable to do this. Clearly, however, the root of the problem goes back to the desire for budget assurances, which NASA is not capable of giving, regardless of what form of cooperative agreement might be used.

In the ISPM debacle, NASA made many difficult choices. Some were choices with little leeway: for example where to cut within each of the categories mandated by OMB, including space science. There was really little room for tradeoff or flexibility within those constraints. In space science, to achieve the OMB target figure an international project was certain to be affected, and it was up to NASA to decide which one.

Certainly the biggest mistake U.S. officials made with regard to the actual handling of the situation was having no prior consultation with ESA officials before the budget was announced. Ordinarily, leaks regarding the budget are the norm rather than the exception in Washington, even with embargo orders. This was not the case in 1981. The demand for secrecy imposed by the White House has been described as near to fanatical. Because, however, the imposed budget cuts would directly impact ESA, the administration, perhaps through the State Department, should have had some sort of consultation with ESA officials prior to the decision being made to cut the U.S. spacecraft in ISPM. This was not the mindset though: the administration demand for secrecy and the overall tendency in Washington to think of international participants in space projects as peripheral prevailed. Had such consultation occurred, much of the angered tone of the diplomatic brouhaha that resulted might have been avoided. Rhetoric to protest the cancellation would certainly have still occurred, as part of a necessary and expected response, but not in the tone of betrayal which was prevalent in 1981 and 1982.

Erik Quistgaard, a Dane, had been appointed ESA director-general in May 1980. His predecessor, Roy Gibson from the United Kingdom, had sometimes been accused of being too pro-American, with critics citing the "unfair" terms of the U.S.–ESA Spacelab agreement. After the fact, many Europeans were unhappy with that agreement. At the time of the signing of the Spacelab MOU, however, many Europeans were grateful to the United States for providing an opportunity for entry into the manned spaceflight program, considering Europe's own less-than-spectacular launch attempt record. In any case, the European program had matured considerably by 1981 and expected to be taken seriously. Therefore, Quistgaard felt a certain amount of personal pressure not to appear to be acquiescing to the unilateral decision of the United States concerning ISPM. Not consulting with Quistgaard prior to the budget announcement gave him little choice in regard to the tone of his response.

Ramifications

Erik Quistgaard made a speech on June 25, 1981, to the ESA Science Program Committee which rather succinctly summarized ESA's post-ISPM attitude regarding cooperation with NASA on future ventures.

> You may well have seen in the press recently, a statement by the new Administrator of NASA that the difficulties with the ISPM project will not have an effect on Europe's wish to continue collaborating with NASA in the space research field. This is, unfortunately, a total misrepresentation of my words, and I would like to take this opportunity to ask the European scientific and technological community not to express too much eagerness for cooperative ventures with the U.S. until the ISPM problem has been solved.[25]

Although the problem was never solved to Europe's liking, cooperation, of course, has continued. The diplomatic waters have been calmed and the European desire for continued cooperative space research has so far outweighed its justified anger over ISPM. Europeans have also used the ISPM situation as a means of driving home certain general points about expectations concerning cooperation on future space projects.

Two projects where ISPM has had a particular impact have been in the space station negotiations and in regard to joint cooperation between NASA–ESA on solar terrestrial science. Concerning the latter, "At least in the early phases of our negotiations with ESA concerning this program, the ISPM debacle was mentioned many times, indicative of the obvious nervousness which they still feel about entering into another joint program with us."[26] ESA wants certain guarantees written into future MOUs in order to prevent unilateral cancellation of such programs or at least to limit its financial liability in the event of cancellation. This is not surprising in view of statements made by Quistgaard and other ESA officials at the time of the U.S. spacecraft cancellation that "Either the MOU becomes more binding—and therefore no longer a gentlemen's agreement—or we will need to restructure the kind of cooperation that we seek in the future."[27]

The problem again, however, is not with the instruments used for international agreements, but with the tenuousness of the political processes in the United States generally, and the budget process specifically. So what can be done to prevent another ISPM incident? First and foremost, the uniqueness of circumstances in 1981 make a reoccurrence highly unlikely. Additionally, former NASA Director of International Affairs Kenneth Pedersen suggested in 1985 that there are certain assurances that NASA can offer ESA, or other partners: "(1) the ISPM experience, with its repercussions, makes it less likely that similar events will take place in the future; (2) budgetary austerity means that only top priority (and therefore not easily cancellable) science projects will be approved."[28]

Certainly NASA is being more realistic about its "new start" expectations. In FY 1979, when ISPM was initiated, NASA had requested five new start programs. Now, OMB and NASA have an informal understanding that space science cannot realistically expect to get more than one new start approved per year. For tactical reasons, sometimes more are requested, but this is primarily to draw attention to the others for priority positioning in subsequent years.

High-level commitment in the United States is also being sought by countries considering cooperative space projects with NASA. Early endorsement of the space station by President Reagan is not being overlooked by potential station participants as an indication of U.S. commitment to the project. Yet endorsement by one president does not guarantee continued commitment from future administrations, a fact other countries are well aware of. Also, the invitation made by President Reagan in his January 25, 1984, State of the Union Address for friends and allies to participate in the development and use of the space station makes international participation optional for both potential participants and the United States. It leaves the type and degree of participation the United States would accept as very nebulous—not necessarily a partnership. There is no real presidential mandate for an *international* space station.

It should also be remembered that congressional attitudes widely vary toward the value and desirability of international cooperation on space projects, with some members strongly objecting to too much international participation or involvement in what they see as primarily U.S. ventures.

Conclusion

The 1981 decision to cut the U.S. spacecraft from the ISPM mission was a poorly handled incident. If giving no prior notice to ESA was the biggest mistake, and it was, then allowing ESA false hope of recovering a two-spacecraft mission after the fact comes in a close second. In all fairness, however, NASA did not knowingly mislead ESA. Rather, ESA and NASA officials initially seem to have fed off each others' bits of encouragement and optimism, with, ironically, NASA eventually taking cues from ESA regarding the status of the negotiations within the U.S. government, since ESA was working with all the parties involved. Certain members of Congress and some State Department officials were particularly encouraging to Quistgaard; this encouragement was probably based on little more than the hope.

Congress took a "cheap shot" at NASA, criticizing it for targeting ISPM but being unwilling to put money back in the budget, as it well could have. Instead, Congress chose to offer ESA rhetorical support and NASA the "option" of reconsidering reinstatement of a second spacecraft in the FY 1982 budget. Of course exercising that option would have been clearly without OMB support for additional money in the future, and thus impossible. In a different political climate, Congress might have been willing to have a high-level confrontation with the administration for more money for space science, but not in 1981.

ESA had just cause for its angry reaction after the United States cancelled construction and deployment of its spacecraft. But ESA was also not beyond making the most of the situation from a political standpoint, attempting to make the United States feel as guilty as possible about its actions as a bargaining chip for ESA in future space project negotiations. ESA may have been naive regarding its perception of the legal status of an MOU, and the U.S. view on the sanctity of international space agreements in 1981, but that is no longer the case.

Endnotes

1. An earlier version of this chapter appeared as an article in *Space Policy*, February 1987, pp. 24–37.
2. Marcia S. Smith, "America's International Space Activities," *Society*, Jan/Feb 1984, p. 21.
3. U.S. House of Representatives, Committee on Appropriations, *Supplemental Appropriations and Rescission Bill*, 1980, 96th Congress, 2d Session, Report #96-1086, pp. 95–96.
4. Irwin B. Arieff, "House Votes $16.1 Billion in Emergency Funds," *Congressional Quarterly Weekly Report*, Vol. 38, No. 25, 21 June 1980, p. 1710.
5. John W. Ellwood, "The Great Exception: Congressional Budget Process in an Age of Decentralization," *Congress Reconsidered*, Third Edition, Lawrence Dodd and Bruce Oppenheimer, eds., Congressional Quarterly Press, 1985, p. 331.
6. Gregg Gail, "Final Carter Budget Sets Challenge for Reagan," *Congressional Quarterly Weekly Review*, 17 January 1981, p. 123.
7. Telex from Erik Quistgaard to ESA Member State delegations, 23 February 1981.
8. Michael A. G. Michaud, *Reaching for the High Frontier* (New York: Praeger Publishers, 1986), p. 200. Michaud also states that an attempt to kill the JOP in the summer of 1977 was the turning point for involvement of the planetary scientists' lobby. When 1,200 jobs at JPL became jeopardized, planetary scientists launched a major telephone and mail lobbying effort, and got others to work on their behalf as well.
9. Letter from J. David Bohlin to Joan Johnson-Freese, 9 February 1988.
10. Telex from Erik Quistgaard to ESA Member State delegations, 24 February 1981.
11. Ibid.
12. Jeffrey M. Lenorovitz, "ESA Seeks Reinstatement of NASA Solar-Polar Effort," *Aviation Week & Space Technology*, 2 March 1981, p. 22.
13. *Aide Memoire*, from the European Space Agency to the U.S. Department of State, 3 March 1981.
14. Telex from Erik Quistgaard to ESA Member State delegations, 17 March 1981.
15. Telex from Erik Quistgaard to ESA Member State delegations, 31 March 1981.
16. Telex from Erik Quistgaard to ESA Member State delegations, 23 April 1981.
17. Ibid.
18. Ibid.
19. Letter to the ambassadors of ESA Member States from Erik Quistgaard, 18 May 1981.
20. Letter from David Stockman to Alan Lovelace, received 22 June 1981.
21. Telex from Erik Quistgaard to the ESA Council and Science Program Committee delegates, 9 September 1981.
22. Message from NASA Administrator James Beggs to ESA D-G Quistgaard, through 1 October 1981 letter to Mrs. Karin Barbance, International Affairs, ESA, Paris.
23. Message from NASA Administrator James Beggs to ESA D-G Erik Quistgaard, through 1 October 1981 letter to Mrs. Karin Barbance, International Affairs, ESA, Paris.
24. See, for example: "Cooperation and Competition on the Path to Fusion Energy," National Research Council, Washington, D.C., Energy Engineering Board, 1984.
25. Director general's speech in reply to the address of the Italian minister of scientific and technological research— SPC, Rome, ESA/SPC/MIN/27, 25 June 1981.
26. Letter from J. David Bohlin to Joan Johnson-Freese, 9 February 1988.
27. "ESA Considers Options for Solar-Polar Mission," *AWST*, 28 September 1981, p. 26.
28. *International Cooperation and Competition in Civilian Space Activities*, Office of Technology Assessment, U.S. Congress, July 1985, OTA-ISC-239, p. 385.

Chapter 8

When the Shuttle Began Flying: A New Emphasis on Commercial Potential

Cooperation Versus Competition

Working under the definition that cooperation is a way to increase benefits to parties jointly involved in some type of venture, by the time the Shuttle started flying it became relatively easy to look back and see how different countries had benefited from cooperation in space. The United States had reaped the benefits of cooperation initially in terms of good will, prestige, and increased stability of programs (most of the time), and later in additional hardware for the Shuttle. Other countries gained an enhanced technological capability and were therefore able to become more independent in their own space efforts. This was particularly the case when the European-developed Ariane family of launchers became operational.

The ability of other countries to act independently of the United States came about at approximately the same time that U.S. policy began moving toward a commercial focus. In at least some instances the United States found itself in competition with those countries for which it had originally served as a tutor. The question as to how to simultaneously cooperate and compete with other countries—or whether or not that was possible or desirable—began to be raised with the advent of space commercialization.

The Reagan Philosophy

On July 4, 1982, President Reagan made the expansion of private, rather than government, investment and involvement in space a major objective of the United States through his National Space Policy. On July 20, 1984, the president issued a National Policy on the Commercial Use of Space. Included among the legal and regulatory initiatives was the provision to assure fair international competition. But the distance between intent and actuality is sometimes great, and in all fairness any international aspects of U.S. commercial space policy soon became almost an afterthought to other problems.

First, it is important to correctly define and differentiate between the terms *privatization* and *commercialization*. Privatization basically refers to turning over tasks, responsibilities, and even facilities held by the government to the private sector. For the most part, space R&D, and indeed space activity and future planning, has traditionally been a prerogative held exclusively by the federal government. In his 1982 speech, Reagan was calling for both privatization and commercialization, where with privatization the private sector would begin to take over in a field where the government had provided what hopefully would come to be viewed as seed money. Clearly, this is a necessary next step for space activity in a capitatlist system such as in the United States, just as it had been some years prior for the airlines. Commercialization, however, basically refers to new, privately funded, usually high-risk initiatives, with profit-making potential. This is a separate, but related matter to privatization, in that privatization can hopefully lead to further opportunities for commercialization. It was commercialization that initially received the most attention, as typically, the cart was put before the horse in a quest for profits—preferably to commence immediately.

Following the traditional pattern of executive initiative and congressional response, Congress amended the NASA Space Act of 1958 directing NASA to "seek and encourage, to the maximum extent possible, the fullest commercial use of space." NASA responded with its own Commercial Use of Space Policy in October 1984, which included the creation of a NASA Commercial Space Office as a focal point for commercial space matters. Initiatives were outlined to stimulate investment in space ventures including measures to reduce financial, technical, and institutional (lessening bureaucratic requirements) risks to U.S. companies. The NASA approach was to work closely with and encourage U.S. private industry in commercializing space. The new policies were designed to benefit U.S. companies exclusively, with the idea that by starting its space commerce program early, the United States could stay ahead of foreign competition and strengthen its position in international trade.

Taking a desired—and even desirable—political goal and turning it into reality is not an easy step. The potential problems that would be encountered with space commercialization were recognized by some analysts early on.

> President Reagan's enthusiasm for the commercialization of space seems boundless: not only was it a major factor in his endorsement of the National Aeronautics and Space Administration's (NASA's) space station, but he has made it a centerpiece of his space policy.
>
> In the real world of politics, money, and bureaucracy, however, it has not been so easy to turn that enthusiasm into a coherent program. The biggest bone of contention at the moment is the pricing of launches on the space shuttle, and the near impossibility of promoting all forms of space commerce simultaneously. But looming in the near future are issues such as international and private sector participation in the space station, the eventual privatization of the space shuttle, and the question of who will call the shots on federal space policy.[1]

Even in the pre-*Challenger* days of NASA and Reagan administration enthusiasm for space commerce there was no guarantee that efforts regarding the commercial use of space would pay off. The impetus for the push toward commercialization was the multi-billion dollar commercial communication satellite business. Although that figure and other estimates of potential space revenue were often cited as inducements, the truth was and remains that space research and manufacturing is a long-term investment that many companies have felt they cannot afford. NASA recognized this and its space commercialization policy was designed to help alleviate the risk to commercial space entrepreneurs who could not otherwise afford to become involved with commercial space ventures due to inexperience in the space field and the long lead times before they would show a return on their investment.[2]

Some firms did recognize space commercialization as a long-term venture and still chose to become involved. For example 3M, a research oriented company, moved into space commercialization early and ambitiously relative to others; in 1985 they had a proposal for an agreement with NASA calling for activity on seventy-two Shuttle flights over a ten year period. Yet even they targeted only a small portion of their total research budget for space after the big 1984 government push.[3] Dr. Les Krogh, 3M's vice-president for research & development, stated, "We foster new science and look at paybacks that may not occur for over 10 to 15 years. We are willing to put in patient money and hard work over a long period of time."[4]

It is important to recognize that much of the work done by 3M in space has been geared to using the knowledge they gained to improve manufacturing processes on the ground, rather than to actually do future manufacturing in space. The same is true for the McDonnell Douglas Corporation and the John Deere Corporation, two other companies which developed programs involving space research. McDonnell Douglas has been interested in electrophoresis, a materials separation process which can already be done on Earth, but can be done in space in a way that makes it possible to increase the volume of the activity by a factor of seven hundred. John Deere Corporation, a farm equipment company, has been interested in the casting aspects of metals and has done a series of experiments in space in that regard.

In the excitement of the moment and in support of commercialization, some advocates seemingly got carried away in their predictions of potential resources and often forgot to mention the probable time lag between investment and return. These fueled almost certainly unreachable expectations for some individuals and companies. One projection concerning potential space revenue estimated a gross figure of $65 billion annually by the year 2000.[5] The sources of that revenue that were most often stressed in 1984 as potentially profitable were: materials processing, including pharmaceutical production ($27 billion), gallium-arsenide semiconductor processing ($3.1 billion), space processed glasses ($11.5 billion); advanced space communications ($15 billion); remote sensing ($2 billion); space transportation ($1 billion); and aerospace support activities ($3.7 billion). The ambitious nature of these figures is clear in that they represent an expectation of an approximately thirty-fold increase in revenues over sixteen years, from the time they were made in 1984 until the year 2000. Another assessment, prepared by Rockwell International, gave forecasts for the commercial space market in the 1990s. That assessment cited a potential $20 billion pharmaceutical market, a $6 billion electronic materials market, and a $4 billion glass market.[6]

Analysts began to question and revise these high figures even before the *Challenger* accident. By 1985 estimates for the projected materials processing market had already dropped to the range of $2.0 to $17.9 billion. This is still a substantial amount, but it does reflect the overzealous nature of previous estimates and should have perhaps injected a note of caution into future planning. But the important point is that President Reagan was a believer. The emphasis put on commercialization by the Reagan administration at least gave the impression that space was quickly going to be a great place to make money, and subsequently companies and countries alike began to posture themselves toward this goal. A look at some of the areas cited as potential revenue sources illustrates the different factors involved and problems encountered during initial commercialization efforts.

Communication Satellites

This is the one area of commercial space activity that is considered established and competitive and where growth is assumed. It was already over a billion dollar per year market in 1984. Because of this, telecommunications is often not ever included in discussions of space commercialization, but considered an entity unto itself. On the other hand, including potential revenue figures from this field in projections only helped the cause of commercialization.

In-Orbit Services

With the advent of the use of the Shuttle, and the "routine access to space" that it was to guarantee, several private firms responded to the belief that there would be a market for facilities which could provide in-orbit services: power sources, communications capabilities, and pointing and stabilization for multiple payloads. This belief was based on the premise that without these facilities, individual payloads would have to bring all of these requisites with them, thereby increasing the cost. A facility would be built, taken into orbit with payloads, and then left there to be serviced on an occasional visit by the Shuttle to pick up, deliver, and perhaps service payloads, as well as service the facility itself. The Shuttle Pallet Satellite (SPAS) platform, a privately developed platform by the German firm of Messerschmitt-Bolkow-Blohm, is an example of one such facility which reached fruition, making its initial flight on STS-7 in June 1983. In 1983 a U.S. firm, Fairchild Industries, announced plans for a "Leasecraft" platform, intended to provide a variety of orbital utilities for a variety of payloads. The initial launch was originally scheduled for 1987, with an estimated cost of $200 million. But by October 1985 it was clear that Fairchild was facing problems due to a lack of commercial customers and difficulty negotiating a service contract with NASA. They therefore put an indefinite hold on the Leasecraft development plans.

Remote Sensing

An area consistently receiving a substantial amount of attention regarding commercial space activities is remote sensing. This basically refers to taking images of the Earth from space for a variety of purposes ranging from resource location to weather predictions. The United States initially led the world in the development of remote sensing systems, primarily through the Landsat program, but would later receive substantial competition from other countries. The history of the U.S. Landsat program and attempts at privatization is well documented and need not be reiterated.[7]

Materials Processing

It is within this field of materials processing that 3M, John Deere, and McDonnell Douglas have primarily focused their efforts. The relatively gravity-free environment of space can reduce the adverse effects of sedimentation, convection, and interaction with container walls in the production of certain products. Pharmaceutical products can be purified, crystal defects can be eliminated, and structures of low density which would collapse in production because of gravity on Earth, can be created in space. Along with pharmaceuticals, semiconductors and glasses are also frequently cited as potential products to be manufactured in space. Even more frequently though, the prospect of magnificent new discoveries and subsequently new products and new industries—with corresponding profits—dazzled the imagination of some, but too often only escalated the estimates of potential revenues that could be had in commercial space fields.

Space Transportation

Until the entrance of the European built Ariane launcher in May 1984 (this flight, Ariane 9, was the first flight for which Arianespace was responsible), the United States had a monopoly in commercial launch services. When the United States terminated its use of expendable launch vehicles (for civilian launches) in favor of the Shuttle, the competition between the Shuttle and the Ariane was widely publicized. However, neither vehicle had any real difficulty keeping its flight manifests full. This led to speculation as to whether launch demands could be met by either or both vehicles in the future. After the *Challenger* accident that question became moot.

International Aspects of the Commercial Emphasis

The European response to the U.S. push for the commercial use of space was directed toward two issues: (1) the role NASA should have generally and (2) the application of U.S. law to foreign companies. First, it was strongly felt among many circles in Europe that NASA's job is research, development, and evaluation of aeronautic and space technology—and that the commercialization of new technological advances, e.g., investment, operations and worldwide marketing, should be left to private industry, much as has been done with Arianespace, the company which is responsible for the production, marketing, and sales of the Ariane rocket. As part of this perspective, it was felt that the Space Shuttle system should be turned over to a private corporation also. Subsequently this might mean that Kennedy Space Center would be "spun off" with such privatization. That would then leave NASA to research as, this perspective says, it was intended—

citing its charter and its predecessor National Advisory Committee for Aeronautics (NACA) as a model and evidence of intent. Without that R&D focus, many people felt that NASA was being pulled in too many different directions, sometimes at cross-purposes.

This view of NASA is not uniquely foreign. In reference to the Landsat program, in 1979 President Carter issued Presidential Directive 54, which included plans for Landsat to become operational and turned over to private industry. Analyst Pamela Mack says of this action:

> Pressing NASA to stay in its role as a research and development agency, not an operational agency, Presidential Decision 54 turned the operational responsibility for Landsat over to the National Oceanic and Atmospheric Administration, which already managed the successful operational weather satellite program.[8]

So as late as 1979, NASA was in fact being pushed to stick to R&D, only to have its entire mission redefined first in 1981 to increase Shuttle operations and then further in 1984 to push commercialization.

Some NASA officials clearly also recognized the problems inherent in being both an operations agency as well as a research agency. Administrator James M. Beggs stated in 1984 that NASA was looking toward 1988, after the Shuttle developmental phase was complete, for privatization.[9] Such, however, was not to be the case. After conferring with the Department of Defense and the White House, NASA announced that Shuttle privatization was not feasible for the foreseeable future. Included in the reasons given for the decision were need for government control of military missions, and national pride in the space program.[10] With regard to commercialization, the change of policy was dictated by the White House, thereby mandating NASA to take on this new responsibility.

On the second issue, many people abroad felt that if the commercial use of space were to be handled by NASA and given a high policy priority, and since foreign payloads were supposedly recognized as financially essential, more care should have been taken to accommodate and encourage foreign users. There were problems regarding the application of U.S. law to foreign companies soon after the commercialization push began, which later overshadowed other negotiations concerning foreign business ventures in space, including the space station. An example of the type of problems that were encountered and that concerned foreign investors, is found in the case of the SPARX Corporation.

SPARX

SPARX[11] was originally to be a joint U.S.–West German enterprise, where the major U.S. partner would be COMSAT and the major West German partner

Messerschmitt-Bolkow-Blohn (MBB). The idea behind the venture was to use the Modular Opto-Electronic Multispectral Scanner (MOMS) developed by MBB, and the Shuttle Pallet Satellite (SPAS) platform, to do remote sensing from the Shuttle on a regular basis and to make the results available for commercial sale. The scanner had already flown on two experimental Shuttle missions in June 1983 and February 1984.

Originally, the corporation was applauded by NASA for their ingenuity and paraded before U.S. private firms as an example of what could be done regarding the commercial use of space. *Aviation Week & Space Technology* (AWST) stated on March 12, 1984, that "McDonnell Douglas with its eletrophoresis biological processing operations and the SPARX Co. effort in commercial remote sensing are the closest to a limited commercial operation in space for a large enterprise other than communications satellites."[12] But the supposedly firm deal (payment had already been accepted by NASA to reserve Shuttle space) was aborted when the Commerce Department threw in a legal monkeywrench by pointing out to NASA that the commercial intentions of the SPARX project would violate the "Open Skies" policy under consideration. The official government stance was, based on the administration's position on the Land Remote Sensing Commercialization Act of 1984[13] which was then being considered in Congress (and was eventually adopted), that all data gathered by sensors on U.S. manufactured spacecraft must be made available on a nondiscriminatory basis and at minimal cost. This virtually quashed the commercial value of SPARX. Further, it raised broader questions on how U.S. domestic law would affect foreign Shuttle—and space station—users and partners. Companies based outside the United States that are interested in utilizing the Shuttle or the space station for commercial purposes, as they are being urged to do by NASA, clearly resent being made the subject of legislation over which neither they nor their governments have any control or influence.

Specifically, it was not clear whether or not the decision applied only to joint missions. Could a foreign company acting alone buy commercial space on the Shuttle and be prohibited from selling its product derived or produced as a result of such because of U.S. law? Certainly Europeans and other potential international space station customers could not help but wonder and worry. In response to a request from NASA to clarify this point, the Commerce Department general counsel stated in a letter that "the requirement to obtain a license for remote sensing activities appeared to be limited to 'those parties otherwise subject to the jurisdiction of the United States.' "[14] Although NASA welcomed this interpretation as favorable to foreign companies, Europeans fears about other potential problems working with the United States were not quashed.

Spacelab

Another situation of concern to many Europeans centered around use of the Spacelab that they had built. When ESRO/ESA signed the MOU regarding Spacelab, they knew what they were getting into regarding their status as a paying customer on all but the first flight of Spacelab. What they did not have was an accurate idea of what was to be the customer cost of using Spacelab. Original pricing estimates were based on multiple annual flights. In 1983, the first Spacelab flight occurred. It was to be followed by one in 1984, but that flight was postponed. The second flight of Spacelab (still designated the Spacelab 3 mission) was in April 1985, scheduled to be followed by two more flights in 1985. Clearly the difference in the number of estimated versus actual flgihts of Spacelab had much to do with the sharp decrease in estimated Shuttle flights: from over 55 in the first five years to less than half that figure. Nevertheless, the difference between estimated Spacelab use and actual use was driving cost up. Why Spacelab was being used even relatively less than originally planned can be debated. NASA claimed it was being used as intended, e.g., when requested and/or needed. Some Europeans took another view: that NASA was increasingly turning from aeronautical and basic space research, which they felt NASA should be focusing on, to making money through the commercial use of space, which NASA clearly was.

Exclusive U.S. Commercial Incentives

Revenue from the commercial use of space has long been a consideration for the Shuttle. It had originally been intended that the Shuttle would be virtually self-supporting through customer payments. This turned out not to be the case.

In a 1975 AWST article it was stated, "A significant influence on NASA ability to maintain viable programs . . . will be the extent to which *commercial and foreign government users* (emphasis added) participate in the development and financial support of future payloads."[15] The problem that developed, however, was that providing exclusive incentives for U.S. commercial users had an inverse effect on encouraging others.

Conclusion

The dilemma over favoring U.S. commercial payloads versus foreign payloads on the Shuttle became moot after the *Challenger* accident. What is important to recognize from these early commercial experiences is that competition was entering into the international space arena to play a role—somehow—along with co-

operation. This was a situation difficult for the United States to accept and adjust to, as it was so clearly foreign to our past tradition of dominance and monopolies.

Unfortunately, privatization and commercialization were often merged into meaning the same thing. U.S. policies, as represented by the increasingly pluralistic spreading around of responsibilities in the field of space within the U.S. government, were well intended but often at odds with each other and poorly implemented.

Endnotes

1. M. Mitchell Waldrop, "Space Commerce: The Quest for Coherence," *Science*, 24 August 1984, p. 812.
2. An explanation of the techniques and mechanisms developed by NASA toward that goal are reviewed in an article by Ray A. Williamson, "The Industrialization of Space, Prospects and Barriers," *The Exploitation of Space*, M. Schwartz & P. Stares, eds. (London, Butterworth & Company, Ltd, 1985) pp. 64–76.
3. See: John M. Logsdon, "Space Commercialization: How Soon the Payoffs?" *Futures* (February 1984):71–78.
4. Judy Schuster, "The Sky's the Limit!" *3M Today*, January 1985, p. 7.
5. Center for Space Policy, Cambridge, Mass. Referenced by Craig Covault, "Unique Products, New Technology," *Aviation Week & Space Technology*, 25 June 1984, p. 40.
6. Ibid., pp. 40–41. See also: Maxine Pollack, "Industry Stakes a New Claim in Space," *The New York Times*, 24 June 1984, pp. 3:1, 26.
7. See: John Logsdon & Tracie Monk, "Remote Sensing from Space: A Continuing Legal and Policy Issue," *Annals of Air & Space Law*, Vol VIII, 1983, pp. 409–431; M. Mitchell Waldrop, "What Price Privatizing Landsat," *Science*, February 11, 1983, pp. 752–754; Pamela E. Mack, "Commercialization, International Competition, and the Public Good," *International Space Policy*, Papp & McIntyre, eds. (New York: Quorum Books, 1987), pp. 195–202.
8. Pamela E. Mack, "Commercialization, International Competition, and the Public Good," *International Space Policy*, Papp and McIntyre, eds. (New York: Quorum Books, 1987), p. 197.
9. See: Waldrop, "Space Commerce: The Quest for Coherence," p. 813.
10. Craig Covault, "NASA, Defense Department Drop Idea of Private Shuttle Management," *AWST* (29 April 1985):42–44.
11. See: David Dickson, "SPARX Fly Over U.S.–German Space Venture, *Science* (8 February 1985):617–618; Dr. G. Greger, "Europaische Erfahrungen & Perspektiven," paper presented at the Internationales Kolloquium Weltraumstationen, Hamburg, FRG, October 1984.
12. *AWST*, 12 March 1984, 99.
13. P.L. 98–365
14. Dickson, "SPARX Fly Over U.S.–German Space Venture," p. 617.
15. *AWST*, 17 March 1975, 59.

PART III

Multiple Actors Give Rise to Choices: 1985–Current

The arguments in favor of cooperation have not basically changed since 1958; there are still political, scientific, and economic benefits to be reaped. Politically, cooperation strengthens the Western alliance and shows the positive aspects of technology. In the early 1980s when "technology" was often equated with the placement of Pershing II and cruise missiles in Europe, and hence had a negative connotation, space technology could be offered as the positive counterpart. In a 1983 article, during a time when alliance strains were evident, Dr. Hermann Strub, head of aerospace programs at the West German Federal Ministry of Research and Technology in Bonn, commented that "Space is something that people can be proud of; it's not attacked in the same way that other technologies are attacked. Cooperation between western countries would be very helpful in bringing everyone together."[1] The importance of political benefits offered from space cooperation should not be underestimated.

But the times are changing. The United States and the Soviet Union are no longer the only players in space with launch capabilities or experience in manned programs. Cooperation no longer is limited to short-term development programs or one country launching a satellite for another. And perhaps most importantly, cooperation clearly must now take place in an environment where competition is also present. These factors have resulted in a new space regime for future activities (Chapter 9).

The Soviet Union recently has begun offering commercial satellite launch services and has actively sought Western participation on cooperative space science projects. Reality dictates that countries with limited space science budgets would be foolish not to diversify their work options. The French experience of maximizing their opportunities by working with both the United States and the Soviet Union shows the benefits of such a strategy. This opening of the Soviet space program to the West will have important implications for the future of international cooperative ventures (Chapter 10).

For both the Reagan and Bush administrations, leadership has been a key objective for cooperative space activity. Yet it is the Soviet Union that is now aggressively using space as an important element in its foreign policy. The Soviets are making attractive cooperative offers regarding both commercial and scientific projects, offers that the United States would have made in earlier times. As to whether the presence of the Soviet space station Mir, in direct contrast to the obvious absence of a similar Western station, and the Soviet ability to make these attractive offers have leadership implications, one analyst has summed it up with an emphatic "Hell yes!"

The space regime now has numerous players with varying degrees of capabilities (Chapter 11). That other countries have matured technologically to a point where they can actively participate in cooperative space programs has been both a benefit and a threat to the United States. Technological maturity inherently brings with it the ability to compete with other countries in selected areas, particularly those with commercial potential (Chapter 12). This has meant that individuals of a mindset geared toward either cooperation or competition, but not both, view this new environment as a dilemma to be dealt with cautiously at best.

The new pluralistic environment of space cooperation, and how the United States has dealt with that environment, has been the setting in which the negotiations for foreign participation on the U.S. Space Station have been conducted. The space station negotiations well illustrate that even with the new environment for space cooperation the protective attitudes of participants in cooperative ventures (Chapter 13) have not changed. It also illustrates the increasing economic importance of cooperation. Cooperation can allow NASA to stretch its own resources; usually not so much by direct project cost-sharing, but more likely by project enhancement through foreign participation, and maintenance costs of long-term projects.

Scientifically, cooperation lately has offered scientists the opportunity to work together to maximize the results to be yielded from their individual projects. Remote sensing, an area of strong interest from many space-faring nations, has been a field of particular cooperative activity (Chapter 14), with a great deal of potential for the future. From the U.S. perspective, cooperation may allow the United States access to data, especially meteorological, from international partners. This benefit of international cooperation was not relevant to the United States until recently.

Cooperation has also afforded U.S. scientists increased research opportunities on foreign spacecraft. Fulfillment of this objective may inherently mean, though, that another cooperation objective, shaping other space programs, is being subjugated. However, many people feel that with the maturing of other space programs, U.S. dominance, not to be equated with leadership potential, will inherently diminish anyway and that the tradeoff—increased scientific opportunities—is not insignificant. Working together on coordination of science missions through groups such as the Inter-Agency Consultative Group (IACG) has shown the payoffs that can result from this type of activity and the IACG has been cited as a model for future cooperative work (Chapter 15).

With economic considerations becoming increasingly important to the future of the U.S. space program, cooperation may well become a means of survival for the U.S. program, as it has been for Europe, Canada, and others. There is little evidence, however, that this premise has yet been recognized by the U.S. government. Certainly this is at least partly attributable to the fact that the United States began with independent capabilities and is now having to learn to accept cooperation as a necessity, as opposed to other countries beginning with cooperative efforts and moving toward autonomous capabilities. A key factor to

be considered here is *why* have these other countries felt compelled to keep with and expand their space efforts. The answer is relatively simple in some cases; in the long term space has served as a way to stimulate their domestic industries. Now it is time for the United States to allow space activity to become a driver in the private sector, as it has been elsewhere for some time, to ease the economic pressures on the United States space program. If becoming involved in more programs means doing the projects on a cooperative basis, then so be it. How far the United States has come in learning to work together, domestically as well as internationally, is the final aspect of analysis to be reviewed, with propositions offered for how to work more effectively in the future by learning from the past (Chapter 16).

Clearly, cooperative activity will have serious implications for which countries are the leaders of the future in space. Increased cooperative activity increases opportunities for leadership. The United States must take heed of suggestions such as that made by the American Institute of Aeronautics and Astronautics (AIAA) to improve our "scientific relationships with foreign organizations" and to "promote more vigorously our scientists' participation in foreign programs."[2] You cannot be a leader if you are not an active player. The key determinant is political commitment: the future of space leadership will depend on who has it and who does not.

Endnotes

1. "After Spacelab, Europe Wants A Better Deal," *Science*, 9 December 1983, pp. 1099–1100.
2. American Institute of Aeronautics & Astronautics, "U.S. Civil Space Program: An AIAA Assessment," March 1987, p. 20.

Chapter 9

The Changing Environment of Cooperation

Changing Types of Cooperation

The nature of international cooperation in space is fundamentally changing. In the past cooperation basically involved either the United States launching something for another country, or, as in the case of Spacelab, cooperation which extended only through the developmental phase of the project. Cooperation now has evolved to include longer and more open-ended programs. These programs, where permanent infrastructures are created and then might be subject to change or modification during their lifetimes, present new problems for international cooperation. Management of open-ended projects may well be the hardest and most contentious point of future cooperation.

The 1986 Report of the National Commission on Space (mandated by the U.S. Congress and appointed by the president) clearly describes a future world where a space infrastructure plays an integral role. The report says:

> U.S. leadership will be based upon a reliable, affordable transportation system and a network of outposts in space. This infrastructure will allow us to extend scientific exploration and to begin the economic development of the vast region stretching from Earth orbit outward to the surface of our Moon, to Mars and its moons, and to accessible asteroids.[1]

The creation of infrastructure inherently carries with it the need for long-term management. Infrastructure of this type is obviously costly, making it reasonable to assume that working cooperatively on these projects will become an increasingly attractive option. On the other hand, the nature of the projects also increases the complexities of working together and the perceived risks of collaboration, especially when "operations must be shared with other political sovereigns over whom only limited control can be exercised."[2]

The obvious example of this new type of open-ended program is the Space Station Freedom. Although NASA will surely retain general control over the station, consultation with other component users/owners is clearly necessary for harmonious operation of the facility. That the environment for space-related cooperation has changed in this regard is evidenced in attorney Nathan Goldman's statement:

> In recognition of the maturing technical ability of Japanese, Canadian, and European space programs, the United States will probably have to share more of the management of the station than it had to share in Spacelab or in the cooperative ventures.[3]

The key words to note there are "probably have to," which suggest that the United States is approaching this situation with a dubious attitude.

First proposed by NASA as a new start in 1974–75 (called the Large Space Telescope), the Hubble Space Telescope (HST) is another, open-ended multinational program, as will be the other parts of the Great Observatories Program.[4] Congress suggested that NASA internationalize the Hubble program because of the potential international interest in its scientific benefits—and because of the high cost of such a project. After more than two years of negotiations, it was agreed that ESA would receive 15 percent of the Hubble observation time in return for ESA providing the solar arrays, the Faint Object Camera, and some staff support for HST operations. The overall management responsibility of the Space Telescope is with the NASA Space Telescope Office, with ESA assigning managers to work with that office concerning their portion of observation time.

Already some technical problems with development have caused some disgruntled feelings on both sides of the partnership. A problem with the solar arrays, involving making the arrays resistant to atomic oxygen after they had already been completed to MOU specifications, has been particularly difficult from a managerial stance. Some NASA managers have felt that as a cooperative partner, ESA assumed full responsibility for certain portions of the projects and therefore should have paid the costs required to make the necessary changes as part of its obligations. At the time the MOU was signed, no one knew that atomic oxygen could be a problem, so it could not have been anticipated in the MOU. When NASA had to assume responsibility for paying the costs of the requisite changes, the NASA relationship with ESA in the HST project became in reality more of a contract relationship than a truly cooperative one.

This situation illustrates an important and increasingly recurring issue: that of the grey area between cooperation and a business relationship, which is sometimes left deliberately ambiguous. As it is a relatively new problem, it has been left unsettled, but over time new definitions will need to be developed and applied to avoid hard feelings between "partners."

Hubble has a life expectancy of fifteen years or more. During that time numerous decisions will have to be made which will certainly affect utilization. Mechanisms and processes must be developed to deal with them to the satisfaction of all involved. ESA, as a cooperative body itself, has some experience with compromise; NASA, in an uncharacteristic position, is behind on this learning curve.

European Capabilities and Attitudes

European commitment to being a major influence in international space activities is shown in their approval for expansion of the 1986 ESA budget of $1.79 billion to $2.24 billion by 1990. This money is in addition to the approximately $900 million (total for all ESA member countries) budgeted by the Member States to their national space programs. Europe has independent and routine access to space through the Ariane family of launchers, a manned-rated space system through Spacelab (making Europe the third space power, behind the Soviets and the United States, to have such a system), a long-term space science program, remote sensing capabilities, extensive microgravity research, and has developed communications satellites used by EUTELSAT and INMARSAT.[5] It is not surprising that Europeans are no longer satisfied with continually taking a subordinate role to NASA.

The goal of European autonomy in space, which the French had supported back to the time of the Symphonie satellite and even before, finally received recognition within ESA in 1985. The ESA Council, meeting at the ministerial level in Rome on 31 January 1985, adopted a Resolution on the European Long-Term Space Plan and Programmes, stating its commitment to maintain and develop European independent capabilities in space and "to prepare autonomous European facilities for the support of man in space."[6] European autonomy was reaffirmed at the ministerial level meeting of the ESA Council in November 1987, specifically citing the objectives that had been set forth and described in the earlier Council, including:

- to prepare autonomous European facilities for the support of man in space, for the transport of equipment and crews and for making use of low earth orbits;
- to enhance international cooperation and in particular aim at a partnership with the United States through a significant participation in an international Space Station.

Although set side by side, some U.S. government officials have had a hard time seeing beyond the word "autonomous" before assuming that Europe has taken a competitive, equated almost with adversarial, posture toward the United States. Hence, European autonomy has been viewed as a negative objective from their perspective.

Clearly, however, Europeans are intent upon autonomy. The Rome objectives go on to state that ESA:

- Considers that an additional effort is needed to ensure that Europe keeps up with other space pow-

Figure 9.1 ESA Ministerial Council resolution. *Photo courtesy of European Space Agency.*

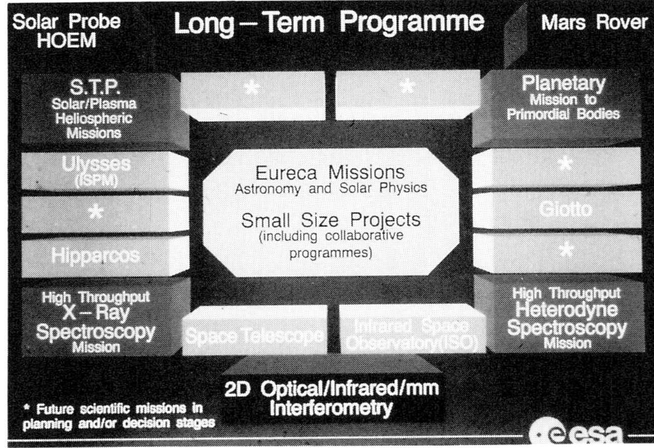

Figure 9.2 ESA long-term programs. *Photo courtesy of European Space Agency.*

ers beyond the year 2000 and to ensure that Europe is capable of all space applications.

- Approves the objective of reinforcing the current European capability in order to achieve as far as possible by the end of this century the capability needed for access to and return from space for manned missions and for servicing payloads, and in order to provide for men living and working in space; Notes the importance of continuing studies and technology programmes concerning future European space transportation systems which will take into account studies carried out *in* Member States nationally and concerning the expansion of the European In-Orbit Infrastructure in order to render it fully autonomous.[7]

After Spacelab and ISPM, the French were able to persuade the Germans to join forces with them in calling for European autonomy, isolating the third major European space power, Britain, which took the position of not wanting to expand to that degree.

It is unfortunate that the Spacelab and ISPM experiences occurred at generally the same time, as it is doubtful they would have provoked the same intensity of feelings separately that they did in combination. Although the Germans were willing to go along with the idea of working toward autonomy, the first step of which was manifested in support for the French spaceplane Hermes, they did insist, however, in continuing to work with the United States also, specifically through ESA support for cooperation on the Space Station.

In July 1984, ESA released its Horizon 2000 program for space science. This program identifies an ambitious space science program for ESA through the year 2000. Attempts were made to incorporate the dichotomous European goals of autonomy and cooperation into this program. This explains some of the ambiguity characterizing recent NASA–ESA interactions and the ambiguity of certain ESA programs, particularly Horizon 2000.

Horizon 2000 is the ESA long-term space science plan, with aspects involving the Solar-Terrestrial Science Program (STSP), x-ray astronomy, sub-millimetre astronomy, and planetary study to include a comet mission. The program proposal includes statements that "The future ESA Scientific Programme must be autonomous, though complementary to other agencies' space science programmes" and "There must be room for desirable cooperative projects with other agencies, although these should not become preponderant."[8] Europe seeks autonomous capabilities, but recognizes that working in isolation is not the most effective scientific course. However, satisfying both goals simultaneously will be difficult and could lead

to some real challenges in the negotiation of an international cooperative program.

Clearly, the current European attitude toward the future is mixed. Although Europeans recognize that cooperation with the United States in the past was a valuable experience, that cooperation on big ticket infrastructure is going to become ever more attractive for budgetary reasons, and that fundamentally they are still not technically capable of carrying out all their ambitions on their own, underlying feelings of doubt and subtle resentment concerning working with the United States remain—sometimes rising up to influence policy.

Another aspect of current European space policy that repeatedly is stressed in policy statements and considered in planning activities is development of national and European industry, both space related and otherwise.[9] This policy aspect, like European autonomy, also takes on certain threatening connotations for some Americans, as it emphasizes the sometimes competitive nature of the relationship between Europe and the United States.

Reimar Luest, director-general of ESA, summarized ESA objectives for the future in a 1987 address in Washington, D.C. He stated them as: (1) Scientific and technical, including: bringing together brainpower, coordinating observations and research efforts, sharing resources for large and ambitious projects, exchanging space opportunities, and stimulating the exchange of advanced technology; (2) Economic, particularly strengthening Western industrial capacity; and (3) Political, encompassing both strengthening transatlantic partnerships and increasing European unity.[10] These objectives reflect that the third period of cooperation between the United States and Europe is one characterized by both cooperation and competition. These objectives reflect that the new environment within which cooperation must take place is one characterized by parallel programs involving both cooperation and competition.

It is important to note in concluding a discussion of European attitudes, that although they present a relatively united front in negotiations outside of Europe, Europeans certainly have their internal difficulties too. In Europe, disagreements are much more likely to be settled internally, without a public airing of their "dirty laundry," than they are in the United States. The United States by tradition is more open about strains within the system. Although this is theoretically healthy, the unfortunate result sometimes is to raise speculations about the certainty of a project or policy and consequently to cause or heighten international tensions.

European plans are ambitious, and some Europeans

are already worrying about their ability to follow through on everything they have planned. ESA is under financial pressure; the degree of pressure is usually commensurate to national election cycles in the member states and the politicians need to respond to national economic situations at the time. This will most likely lead to a pragmatic recognition of reality and subsequently a toning down of autonomy rhetoric. ESA will want to maintain and strengthen, and perhaps diversify, collaborative ties.

Attitudes Within the United States

Americans are used to being preeminent in space. The preeminence which the United States enjoyed for so long created a situation where Americans are used to directing projects and having things their own way. Once you have had the whole pie, it is difficult to adjust to having only part of it—and some people have found it very difficult to make that adjustment. Although the problem may in some respects diminish through the replacement of "old school hard-liners" with new people, the national security, technology transfer, industrial advantage, and scientific opportunity aspects of the problem will clearly remain.

Government

U.S. motivations for cooperation in space activities have evolved as the environment, attitude, and benefits have changed. According to then Acting Deputy Assistant Secretary of State Michael Michaud in a May 1988 speech, the United States does not enter into cooperative agreements for idealistic reasons. Rather, they are entered into for pragmatic reasons.[11] The pragmatic reasons then cited include (1) to increase the scientific and technical return from space activities; (2) increase our (U.S.) ability to influence international practice, international law, and international institutions in the space area,[12] and (3) burden-sharing, in terms of enabling us to do things we could not do within the limits of our own resources.[13]

Some people in the U.S. space community question the value of cooperation if the United States cannot be the managing partner. This might be particularly difficult for NASA people to concede, as some have exhibited the tendency in the past to expect to tell people what to do, like contractors. Such an attitude, however, ignores the reality of the situation, that this is simply no longer possible in every instance. Further, the United States has already taken a secondary role to other countries in space programs and still reaped enormous benefits. A particular case in point concerns the international flotilla of spacecrafts to Halley's Comet in 1986.

NASA participated in the mission through the Inter-Agency Consultative Group, which was the coordinating body for the Halley's activity. (See Chapter 15.) Although the mission was a resounding success, it received relatively little attention in the United States because there was no U.S. spacecraft, and the attention that was drawn to the mission was often negative, again because there was no U.S. spacecraft. Yet, NASA Associate Administrator for Space Science and Applications Burton Edelson, in testimony before Congress, said that he saw the much criticized lack of a U.S. spacecraft in the Halley's mission as a blessing in disguise. Edelson pointed out that it would enhance foreign cooperation on space science projects—which is becoming an increasingly a cooperative realm in order to decrease costs and maximize benefits—and foster "unprecedented" Soviet interest in future cooperation.[14]

NASA Administrator James Fletcher, in a December 1987 statement before the House Subcommittee on Space Science and Applications, succinctly summarized the space world as it is today; the one in which NASA must operate.

The Task Force[15] points out correctly that the space world is a much different place today than it was thirty years ago—or even ten years ago. We have moved from an initial era of two space-faring nations to an era in which those two nations gain a great deal from cooperation with partners who have more recently entered onto the scene. There can be no doubt that the United States has enhanced its own national space program through a history of cooperation with NASA's partners abroad.

International cooperation has become an integral part of the U.S. space program. We agree with the Task Force's assessment that such cooperation will increase substantially by the period 1995–2000. It is important to recognize that this is not a pessimistic scenario. The U.S. is not falling behind; rather other nations are developing their own capabilities. As a result, the prospects for enhanced science and technology are increasingly exciting.

This successful and fruitful cooperation is in many cases with the same countries who now have become our competitors in other fields of space endeavor. In fact, strong foreign capabilities provide for competent foreign partners who can continue to bring significant benefits to NASA and the national through partnership in the more ambitious joint projects. At the same time, however, this strength also makes them more formidable competitors in the commercial arena.

In summary, as we pursue NASA's programs in the future we should keep in mind that in spite of this increasing competition from abroad, competition and cooperation need not be incompatible.[16]

Acknowledging the way of the world and acting in a way consistent with the environment of this world has, however, been a task easier said than done.

The project where the current ambiguity of the U.S. position on international cooperation is most evident is the space station. When Ronald Reagan directed

NASA to "develop, within a decade, a permanently manned space station" in his January 1984 State of the Union address, he issued an invitation to friends and allies of the United States to participate. There was no offer of a joint, international space station; there was an invitation for friends and allies to participate—in some role to be agreed on by both the interested party and the United States—on a U.S. project. Even that invitation was not without debate; some advisors adamantly insisted that the invitation be omitted from the president's speech.

What followed the president's speech was an initial flurry of international activism, and then NASA gave way to reality—reality being that in order for Congress to agree to fund the space station, NASA would have to satisfy national interests by concentrating on the domestic aspects of the space station program.[17] Congress was not always gentle in its prodding of NASA to put domestic interests first. The HUD-Independent Agencies Subcommittee of the House Appropriations Committee is the subcommittee with responsibility for the NASA budget. In a January 1986 letter from Congressmen Edward Boland and Bill Green, the respective chairman and ranking minority member of this subcommittee, to NASA Administrator William Graham it was stated that:

> We want to make it clear that it is the Committee's view that the United States should have the *primary* role in developing the station and should derive the *principal* benefits—particularly in view of the fact that NASA will contribute up to 80% of the cost of the program. Therefore, it would seem reasonable and essential that as issues of foreign participation are clarified—the United States will be awarded the lead role in development of the microgravity module and equipping other modules—be they foreign or domestic.[18]

The position taken here makes sense: to protect American interests as the primary investor in the space station. What it fails to give due consideration to is that potential partners are going to want returns from their investment as well.

Basically, management of multinational, open-ended projects is an area where Europeans have had far more experience than Americans. One of a very few areas where the United States has had extensive experience regarding a multinational infrastructure is NATO, which is for some reason seldom used to draw experience from within the United States. Michael Michaud has said, "I think we [the United States] are going to have to develop a whole set of people scattered around through the appropriate agencies who know how to run multinational programs, and that core of people does not exist at this time."[19] So from a political perspective, cooperation entails not only attitudinal factors but implementation factors as well.

Scientists

Another source of antagonism toward a "spirit of cooperation" has been that U.S. scientists have felt a lack of reciprocity with Europe with regard to provision of opportunities for participation on European scientific missions. Although Announcements of Opportunity (AO) for scientific experiments on NASA missions have traditionally been open to anyone of any nationality, ESA has restricted its mission AOs to European scientists. Specifically, this meant that Americans could be co-investigators on European missions but the principal investigator had to be European. This policy went along with ESA's general policy that members would receive returns commensurate to their financial contributions, and as the United States was not a paying member, participation of U.S. scientists was viewed as taking away from the members' opportunities. In another respect, it was also a kind of affirmative action plan for the European science community, an enhanced opportunity for them to catch up to the U.S. scientific community. But American scientists did not see it that way, especially when they felt that in some cases European experiments were being chosen over U.S. experiments for NASA missions because they were provided free, as opposed to NASA having to pay the costs of American experiments. Recognizing the ill will that this ESA policy was causing, the policy was changed by mutual NASA-ESA agreement in 1983.

Conclusion

Space is now an arena with multiple players having varying degrees of capabilities, and cooperation has taken on a new, long-term meaning. Mechanisms for dealing with this situation are far more familiar to some actors than to others. In addition, the degree of political commitment varies, with uncertainty the unfortunate watchword characterizing the position of the United States.

The United States is no longer able to be the leader in space activity simply through the default of the Soviet Union and other countries. And unfortunately, by default the United States is losing the leadership position it was once left to establish by these other countries. If the United States intends to regain its strong leadership position, it will have to make attitudinal and bureaucratic changes to suit this new environment. Countries are currently posturing to assure themselves an active role in the development of future space technologies. There will clearly be both cooperative and competitive aspects to future activities, making it imperative that they not be viewed as either/or propositions.

Endnotes

1. *Pioneering the Space Frontier*, The report of the National Commission on Space (New York: Bantam Books, 1986), p. 7.
2. Kenneth Pedersen, "The Global Space Context: Changes and Challenges," in Molly K. McCauley, ed., *Economics and Technology in U.S. Space Policy* (National Academy of Engineering, 1987).
3. Nathan C. Goldman, *American Space Law* (Ames, Iowa: Iowa State University Press, 1988), p. 147.
4. The Great Observatories program included the Hubble Space Telescope and the Gamma Ray Observatory, together with the proposed Advanced X-Ray Astrophysics Facility and the Space Infrared Telescope Facility.
5. NASA Advisory Council, Task Force on International Relations in Space, *International Space Policy for the 1990's and Beyond*, October 12, 1987, p. 6.
6. ESA/C-M/LXVII/Res. 1 (Final), 4 February 1985, pp. 1,3.
7. See: *European Space—On Course for the 21st Century*, European Space Agency, BR-39, 1987, p. 9.
8. Ibid.
9. *International Space Policy for the 1990's and Beyond*, NASA Advisory Council, Task Force on International Relations in Space, October 12, 1987, p. 5.
10. Reimar Luest, "The Cooperation of Europe and the United States in Space," Lecture before the Association of American Universities, The Fulbright 40th Anniversary Lecture, 6 April 1987, Washington, D.C., p. 20.
11. Michael A. G. Michaud, "Space Policy at the State Department," *Space Times*, American Astronautical Society, September–October 1988, p. 1.
12. Michaud discusses that the United States has generally rejected the idea of regulation of the use of space, as restrictive of our freedom of action. Clearly, it is not perceived as beneficial to the broadly defined U.S. national interest to expand space regulations, as it was when regulations were established in areas like satellite communications.
13. Michaud, "Space Policy at the State Department," pp. 5,6.
14. "Halley success boosts international missions," *Aerospace America*, 5/86, p. 1.
15. Referring to the NASA Advisory Task Force on International Relations in Space.
16. Dr. James C. Fletcher, Statement before the Subcommittee on Space Science & Applications, Committee on Science, Space and Technology, U.S. House of Representatives, December 10, 1987, pp. 2,5.
17. See: Wulf von Kries, "Flunking on Space Station Cooperation?" *Space Policy*, February 1987, pp. 10–11.
18. Letter from Bill Green and Edward Boland, to William Graham. January 14, 1986.
19. Interview, Michael A. G. Michaud, Washington, D.C., 4 May 1988.

Chapter 10

Glasnost in the Space Arena

After twenty-five years of working behind closed doors, the Soviet space program has opened its doors to the West.[1] The Soviet Union has begun offering commercial satellite launch services, has actively sought Western participation on Soviet space science projects, and has offered to launch the U.S. space station on its heavy-lift launch vehicle. Although it is highly improbable that some of these offers of cooperation will reach fruition, the new Soviet *glasnost* regarding international cooperation in space will clearly be one of several factors affecting patterns of cooperation in the future. Why the Soviets have changed their posture toward cooperation on international space ventures, how the changes have occurred, and the probable consequences of those changes on traditional patterns of international cooperation in space are all relevant questions. They can be answered through examination of the attitude of the Soviet government generally, the benefits they foresaw, and the political environment in which the changes took place.

The United States and The U.S.S.R.: The Primary Actors in Space

Glasnost in the Soviet space program is clearly part of General Secretary and Soviet President Mikhail Gorbachev's broader *glasnost* policy. Had the Soviet Union not been looking to better relations with the West overall, cooperation in space would have been a difficult, if not moot issue. That it has occurred is testimony to Gorbachev's authority, which has been strong enough to allow for change even in areas traditionally dominated by the military, and reflective of the influence of some space advisors, particularly Dr. Roald Sagdeev, former director of the Soviet Space Research Institute (IKI), and currently the principle science advisor to General Secretary Gorbachev. Sagdeev was succeeded in 1988 as director of IKI by the man many people consider his personal choice for the position, Albert A. Galeev.

Recent technical problems involving the Soviet Phobos mission illustrate that their space program is not invulnerable. The technical problems with Phobos basically resulted in the loss of both spacecraft and will most likely lead to a reevaluation of the ambitious nature of the entire program, which has been a drain on the already suffering Soviet economy. Their space station, MIR, was also temporarily mothballed in 1989 due to technical problems, with some political and economic overtones. The internal debate surrounding the appropriate nature and scope of the program has also been reflected in quite open disputes between Soviet space leaders. Still, the Soviets have taken on a new role for themselves internationally in space affairs, one that will continue, albeit perhaps at a slower pace than originally expected.

That the Soviets have indeed embraced an attitude which accepts and even encourages cooperation in space science is clear. In the Yearbook of the Soviet Institute of World Economy and International Relations, *Disarmament and Security 1988*, it is stated that:

> ... In addition to having purely practical advantages the joint work of scientists, engineers and designers, the broad exchange of research findings and the broadening of contacts between specialists and researchers will contribute to the establishment of a favorable political atmosphere through the enhancement of mutual understanding and trust among nations, the elimination of old stereotypes and the removal of excessive suspiciousness and prejudice.
>
> But there are still considerable difficulties in realizing such cooperation. The political issues, secrecy, and problems of the transfer of technology that may have military significance still play a major role in the field of space exploration. Nonetheless, practical experience shows that these obstacles are quite surmountable.[2]

Evidence that the Soviet attitude toward cooperation with the West had changed began to emerge in 1985.

The entrance of the Soviet Union into the international space community came about during a time when, for the most part, the Soviets were "riding a high" in space accomplishments. In February 1986, while the American public was still mourning the *Challenger* tragedy and officials in the United States were lamenting over the task of rebuilding the U.S.

space program, the Soviet Union successfully orbited a new space station, called MIR. That event received relatively little press or public attention in the United States. Elsewhere in the world, however, the event was lauded as a major space milestone. For three years, until its temporary mothballing in April 1989, MIR served as the first permanently manned space station and as the flagship of a well-developed, consistent effort in space by the Soviet Union.

Since approximately 1986 (some analysts cite the 1986 International Astronautical Federation meeting in Innsbruck as the turning point), the Soviet Union has adopted a "new posture" concerning space: more emphasis on the nonmilitary aspects of space; an increased desire to work with Western countries on space science projects; offering their boosters for commercial launches; and an openness atypical of the Soviet Union in the past. It is rather ironic when compared to the activities of the United States during the same period: more emphasis on the military aspects of space, including the increased and open militarization of projects previously thought to be primarily civilian in nature like the Shuttle, as evidenced by the post-*Challenger* payload manifest placing heavy emphasis on military payloads; the difficulties encountered in negotiations with Space Station partners; the Shuttle ceasing to compete for new satellite launch contracts; and a general slowdown of space science projects. Upon examination it becomes apparent that the posture changes in the Soviet space program reflect a pragmatic appraisal of the political environment, including the problems and changes in the U.S. space program, and a logical—and limited—response.

It used to be that comparative pictures of the Soviet and American space programs were drawn simply from adjectives used to describe each. The American space program has always been characterized as exciting, dynamic, one to take on monumental tasks, prestigious, and—to a certain extent—glamorous. The Soviet program, on the other hand, had been described as secretive, rather plodding in nature, incremental in its achievements, one to build on past technology rather than go for spectacular new technologies, and to a large extent, rather "grey" and unsexy.

But it is increasingly recognized that those same qualities can also be cast in a different light. The U.S. space program has been subject to the fits and starts of both public/private interest and funding, plagued by demands for spectacular—and costly—missions to boost U.S. prestige, without any long-term plans to build toward, and subject to ridicule and constant political pressure because its open nature leaves every flaw rawly exposed to criticism from every imaginable source. The Soviet program on the other hand—with the luxury of working behind closed doors except

when the Soviets chose to open them to exhibit their successes—has developed with consistent support and funding, has plodded steadily along building on the technology which has been mastered over time, and has determined long-term future plans. The *Baltimore Sun* ran a cartoon to accompany an article entitled "New Challenges from Allies and Soviet to U.S. Space Supremacy"[3] which shows the NASA hare sleeping under a tree while several tortoises, led by the Soviet Union, make their way by. The unsexy, plodding, incremental Soviet program clearly is not without its strengths.

Past Soviet Cooperation in Space

During the 1960s and 1970s, the Soviet Union's primary partners in international cooperative space ventures were other Eastern-bloc countries. The Apollo-Soyuz linkup of 1975 (see Chapter 6) was the most publicized exception to that general rule, although there have also been others. For example, as part of an information-sharing arrangement in the 1970s, the Soviet Union provided NASA with medical data on the effects of long-term human exposure to weightlessness, which was valuable to early Shuttle crews. Generally, however, the Soviet program was closed to the West, to the extent that the existence of their Plesetsk launch site was not publicly acknowledged by the Kremlin[4] until many years after its location had been determined by Western analysts by backtracking satellites launched from that site.

Within the Eastern bloc, a program called Intercosmos was developed in 1970, involving Bulgaria, Cuba, Czechoslovakia, East Germany, Hungary, Mongolia, Poland, Romania, the Soviet Union, and Vietnam, to promote scientific research projects on space topics on a collaborative basis.[5] These projects were obviously directed by the Soviet Union. As part of the Intercosmos program, crew members from each of these countries have flown on Soviet space missions.

There have also been exceptional agreements outside of the Eastern bloc countries. For example, a limited agreement for scientific cooperation was constituted by an exchange of letters between the U.S.S.R. Academy of Sciences and the European Space Research Organization (ESRO) in 1971.[6] It dealt with the periodic exchange of scientific and technical information published by both without restriction as to use, the exchange of information on scientific programs and projects of mutual interest, and an agreement in principle to explore the idea of the exchange of specialists. The agreement is still valid and has been the basis for contacts between the Russians and the ESRO successor organization, the European Space Agency (ESA).

Of a more specific and extensive nature has been

the cooperation in space ventures between France and the Soviet Union, which began in 1966. The extent of the French–Soviet cooperation is shown by the comparative monetary amounts France has spent on cooperation with both the United States and the U.S.S.R. In 1979 France spent 27.62 million French francs on space cooperation with the U.S.S.R., and in 1980, 32.33 French francs. During those same years, France spent 24.6 million French francs and 38.35 million French francs, respectively, on space cooperation with the United States.[7] The comparability of these figures is reflective of French efforts to maximize their opportunities by pointing out to both the United States and the U.S.S.R. the opportunities offered by the other, with the hope that neither party will allow themselves to be grossly outdone, and so working on about an equal basis with each. French presidents de Gaulle and Pompidou were for many years the only known Westerners allowed to visit the Baikonur Cosmodrome[8] (in 1966 and 1970, respectively), evidencing the rather extraordinary nature of the relationship. The most visible form of cooperation has been the participation of French cosmonaut Jean-Loup Chretien on the Soyuz T-6 mission in 1982, and on the Soviet–French Aragatz mission to MIR, including an ELV for Chretien, in December 1988.

The usually harmonious Soviet–French space relations were somewhat shaken in April 1987 by allegations from the French that the Soviets were trying to obtain Ariane technology. A French scientist working on the Ariane, married to a Soviet woman, was charged with stealing information on the Ariane cryogenic engines for the Soviets. The Soviets called the charges absurd and concerns were voiced about the possibility of further cooperation in the wake of the episode. The ramifications remained confined only to strong rhetoric.

India has also participated in several cooperative space ventures with the Soviet Union, most notably having Rakesh Sharma, a squadron leader in the Indian Air Force, on the Soyuz T-11 launched April 2, 1983. For many years, Indian scientists were among the few foreigners allowed at Soviet launch facilities.

Indications of Change in the Soviet Program

Participation in International Organizations

There have been several indications of change evident in Soviet policy regarding international cooperation in space. For example, in August 1985, the Soviets signed an agreement with INTELSAT for a mutual exchange of technical and operational information.[9] This agreement has been seen as a step toward full INTELSAT membership for the Soviets. This is a rather dramatic change from the past when they denounced the organization as being dominated by Western powers. Other indicators have been equally clear.

Formation of Glavkosmos

In late 1985 the Soviet Union formally established an organization known as Glavkosmos, under the directorship of Alexander Dunayev. At its inception, the purpose of Glavkosmos was stated as being to improve management of the country's civil space programs and streamline the interface between the Soviet Union and other countries for cooperative programs. Until Glavkosmos, the Soviets did not have a national space organization. "Space" was to a large extent subsumed by the military. The "nonmilitary" aspects of space were run by the Intercosmos Council of the Soviet Academy of Sciences.

Space Science Cooperation

In March 1986, Halley's Comet was encountered and surveyed by Soviet, European, and Japanese spacecraft. The Soviet Union in fact led this multinational study and exhibited behavior uncharacteristic of the Soviet Union in the past. That is, Western scientists and journalists were allowed to wander freely around previously restricted facilities, buildings, and corridors. That the Soviets enjoyed their role as host was clear. Also Sagdeev, as the director of the Soviet efforts concerning Halley's Comet, was later summoned to the Kremlin by Gorbachev and decorated with the Order of Lenin and proclaimed a Hero of Socialist Labor.

The candor and direction of the Soviets exhibited at the October 1987 International Space Future Forum in Moscow, hosted by Sagdeev, further encouraged Westerners that a change of attitude had indeed taken place in the Soviet Union. Success in ventures such as Halley's Comet, and these indications of a more open scientific program, have interested Western scientists in further cooperative projects. In some cases, they have even led to larger space cooperation agreements.

The Soviets have been making an overall effort to interest certain countries into closer cooperative ties, efforts which have not been without some success. Representatives of the Intercosmos Council and the British National Space Center (BNSC) met in Moscow in September 1986. As a result of that meeting, a protocol was signed outlining possible areas of further cooperation between Britain and the Soviet Union in space, including radio astronomy, solar terrestrial physics and planetary science, infrared and submillimetre astronomy, space biology and medicine, and materials research. This then led to an intergovernmental agreement on space science cooperation between the U.K.-U.S.S.R., signed during the visit of

British Prime Minister Margaret Thatcher to the Soviet Union in April 1987.[10]

The launch of the first additional module to be attached to MIR, Kvant, also occurred during the visit of Prime Minister Thatcher. Mounted on Kvant is an x-ray telescope developed by Birmingham University, in cooperation with the Soviet Academy of Sciences Institute of Cosmic Research. Realistically, working with the Soviet Union on collaborative space science ventures will significantly increase the opportunities for Britain (or any country) in space science, who in the past dealt exclusively with ESA and with development of European payloads for NASA spacecraft. The new Kvant laboratory carries x-ray telescopes developed by Soviet, British, Dutch, West German and ESA scientists.[11]

Although the Western scientists did not know the exact launch date of Kvant in advance, they knew on a personal basis of a four to six week launch window. They had to deliver their experiments to IKI and could not observe the launch, but they were consulted and informed in advance regarding the switch-on of their experiments. Regarding access to information, in the case of the West German HEXE experiment, for example, the shared observation time is two-thirds Max Planck Institute, one-third IKI. The data is available for all the partners, but evaluation is made by each institute separately.

The Soviets were anxious to conclude the agreement with Britain for both scientific and political reasons. They felt the work with Birmingham could be continued, and that an agreement with Britain would further confirm their membership in the international space community, from which they had long excluded themselves. The British wanted to conclude an agreement only if specific areas of cooperation could be identified, and perhaps would have preferred a less formal agreement than one of an intergovernmental nature. They were not interested in "blanket agreements," with no clear purposes set out. The work between Birmingham and the Soviet Academy of Sciences did provide a foundation for further technical work, allowing scientists to convince the British government the agreement would be worthwhile.

After the signing of the agreement, rumors quickly began that a Brit would fly on a Soyuz mission.[12] British astronaut Nigel Wood had been scheduled to fly on the Shuttle in June 1987. As that was no longer possible, there was speculation that the Soviets might offer a flight to MIR as an alternative, with some obvious public relations benefits. Initially there was little interest in manned spaceflight on the part of Britain though, feeling that their limited space resources could better be spent elsewhere. Further, of the possible astronaut candidates from the U.K., in 1987 all were

military persons whom the British government did not want to send to the U.S.S.R.

In July 1989, however, a commercial agreement was signed for an Anglo-Soviet space mission, called Juno, involving a British astronaut performing scientific experiments on MIR as part of an eight day mission between March and July 1991. The project will not be funded by the British government; funds are to be raised by the sale of sponsorship and merchandising packages, as well as payload space for scientific experiments and broadcasting rights. Approximately $25 million is needed to train and send the British astronaut into space. Two candidates, an astronaut and a backup, will be selected from more than 11,300 applications received. They will then be sent to the Gagarin Center in Star City (near Moscow) for eighteen months of training, including learning to speak Russian.[13]

The United States and the Soviet Union also signed a cooperative agreement in 1987. There had not been such an agreement since the United States allowed the one initiated in 1972 to lapse in 1982 as part of a sanctions package resulting from Soviet activities in Poland.[14] Some individuals and institutions in Washington, significant factions within the Pentagon and the National Security Council in particular, are still against such cooperative agreements. Reports that U.S. intelligence sources believe the Soviet Union saved several hundred million dollars in the R&D of its shuttle program from U.S. technology transfers[15] fueled the fire. NASA termed the assertion groundless, but concerns such as these certainly have been the basis of opposition to expanding U.S./U.S.S.R. cooperation.

U.S. space scientists and space advocates had wanted to maintain the cooperative space agreement all along, rather than linking it to other foreign policy objectives. But it was not until 1984, in an effort to emphasize the peaceful aspects of the U.S. space program and blunt Strategic Defense Initiative (SDI) criticisms, that the political environment changed so that the Reagan administration was willing to again consider a cooperative space agreement with the Soviets. The Soviets, however, recognizing the motivation, rejected Washington's informal overtures on the subject. They, by that time, wanted to link any space agreement to the SDI project.

The first sign that an agreement could perhaps be worked out was during the visit of Soviet cosmonauts Valeriy Kubasov and Aleksey Leonov to Washington in July 1985 to celebrate the tenth anniversary of the Apollo-Soyuz docking. The cosmonauts avoided speaking out on SDI, which U.S. space officials took as an indication that the Soviets might be willing to sign a cooperative agreement not linked to SDI. In the

United States, the release of several Soviet dissidents was seen as an encouraging sign concerning human rights policy in the Soviet Union, enough of a move to let U.S. officials supportive of a cooperative agreement push it both within the Reagan administration and to the Soviets. Shortly after the visit of the cosmonauts, in October 1985, a delegation of U.S. congressmen and NASA officials visited the Soviet Union to try and lay the groundwork for a cooperative agreement. Soviet scientists were interested but noncommittal. For the next several months informal overtures continued, primarily between U.S. and Soviet scientists.

A new U.S./U.S.S.R. cooperative agreement was finally drafted in October 1986, after JPL Director Lew Allen, Jr. led a scientific delegation that held secret meetings in Moscow during September, focusing on sixteen cooperative programs of data exchange and program coordination. The scientists had hoped a joint Mars sample return mission would be included in the agreement, but U.S. officials declined to include it in the final version, saying there were no approved plans for such a mission. Department of Defense (DOD) concerns over technology transfer, and the limited NASA budget were underlying reasons for the omission. The final agreement was signed in April 1987 by U.S. Secretary of State George Schultz and Soviet Foreign Minister Edward Shevardnadze.

The payoffs from the agreement for both parties were primarily political. The United States got a boost for detente; but many Reagan administration officials felt that was insufficient to make the agreement worthwhile. NASA officials, however, supported the agreement as a way to loosen the ever-tightening grip of the military on space projects and funding. The Soviets got the prestige from the appearance of technical equality that inherently accompanies the agreement, as well as close exposure to U.S. space technology and operational procedures—hence the DOD concern.

There has been speculation that the Soviets would also like to sign a cooperative agreement with West Germany. Currently, in cases like the HEXE experiment on Kvant, cooperation is without federal government interaction, but rather on the basis of an agreement between Deutsche Forschunggemeinschaft (DFG) and the Soviet Academy, with a special letter of exchange between the Soviets and the Max Planck Institute (Garching) in 1982.

As of July 7, 1987, a general cover agreement on scientific-technical cooperation between the FRG and the Soviets concluded on July 22, 1986, entered into force. In the general cover, or frame agreement, space is mentioned as a possible field of cooperation which could be developed in a future detailed agreement, and discussions have recently been held in that regard. For the most part, however, FRG-U.S.S.R. cooperation in space has been only on a very limited basis.

Other countries mentioned as being of particular interest to the Soviets for future cooperative space science work include Japan and China. Before the *Challenger* accident, there had been discussion that a Chinese astronaut might fly a Shuttle mission. When that became impossible—as with the case of Nigel Wood—the Soviet Union even hinted at a joint Sino-Soviet Soyuz mission.[16]

Generally, Soviet flight requirements for safety, reliability, and technical perfection of equipment and experiments can be somewhat more lax than those required by NASA for scientific cooperation. Therefore, it can be cheaper, though possibly more risky, for Western scientists to work with the Soviets. Sometimes, for scientists with limited funding, this is an attractive notion even with the possible risk. Soviet interest in attracting Western participation in space science ventures has been such that in at least some cases it is reported that they are offering Western scientists cheaper and more favorable conditions for cooperation than they are for scientists from Eastern bloc countries.

Commercial Satellite Launches

In June 1986, in the wake of launch failures by both the United States and Europe, the Soviet Union announced its intention to enter the commercial competition to place satellites in orbit via expendable launch vehicles (ELVs). The Soviets said that they felt ready to enter the commercial market. An Indian satellite to study natural resources, the fourth in a series of four Indian satellites scheduled for launch by the Soviets, was to be the first carried out on a commercial basis, as opposed to being free to India as in the past.

Glavkosmos promotes several types of commercial space services, including: the launching of satellite payloads, primarily with the Proton booster; vertical sounding rockets; lease of transponder relay capacity on Gorizont communications satellites; sale of Earth resources data collected by Soviet satellites, including photos taken by the KFA-1000 camera which offers the sharpest imagery currently on the market; and use of Soviet developed or customer supplied systems to perform materials processing and other microgravity experiments on Soviet spacecraft.

Glavkosmos officials have mounted an aggressive marketing campaign for Proton use, issuing over 100 solicitations worldwide to embassies and international organizations. The solicitations were particularly timely in that they were first issued just after the first post-*Challenger*-accident U.S. Shuttle manifest was released in October 1986, one which disappointed many Europeans. Space Commerce Corporation, Houston, headed by lawyer Art Dula, is the Proton marketing agent in the United States.[17]

In May 1987 a six-member Soviet space team from Glavkosmos and Licensentorg, a Soviet foreign trade agency, made a three-city tour of the United States to market launch services and materials processing payload space on unmanned Soyuz missions and on MIR. The Soviets displayed an impressive full-size mockup of the MIR space station and Kvant module at the Paris air show in June 1987. They have also conducted marketing trips in France, Japan, and Australia, as well as started looking into the opening subsidiaries in Western countries—one has already been opened in West Germany. Further, talks have been held with Australian officials concerning the possibility of a jointly owned Soviet-Australian company launching Proton rockets from a Queensland space base. They are clearly committed to becoming an active international player in the commercial launch field.

The attraction of the Soviet launch offer to the West is simple; they are offering commercial satellite operators prices of approximately $24 million for a 2.5 ton spacecraft to geostationary orbit. This is about half the cost of a similar Ariane launch. Of course the Soviet government highly subsidizes its launch rates, making these prices possible. This bothers U.S. trade officials who fear U.S. companies will want "a cheap ride . . . to the detriment of the commercial U.S. launch industry."[18]

Arrangements can also be made for payment in technology rather than cash, and not insignificantly, the Soviets have offered assistance in obtaining launch insurance below market rates. They began by working with insurance companies outside the Soviet Union, moving then toward offering space launch insurance for satellites launched from its facilities through Ingosstrakh, Ltd., a Soviet insurance company, another expansion of Soviet commercial activities.

Although the Soviets have no firm contracts yet beyond the Indian launch, in 1986 Glavkosmos went after the contract to be alternate launcher for Eutelsat. The primary launcher is Ariane and the Shuttle was to be the alternate launcher. Under consideration as alternate were the Proton, the Martin Marietta Titan, and the General Dynamics Atlas/Centaur. The price offered Eutelsat for a Proton launch was approximately $26 million in 1985 dollars, to be paid in Swiss francs.[19] Further Glavkosmos Director Dunayev stated that Glavkosmos could integrate and launch an unfamiliar satellite in about 18–20 months, an attractive consideration for potential customers who could wait two or three times that period for an Ariane launch. Eutelsat eventually reserved space on two Atlas rockets in early 1990, and the Atlas/Centaur was selected to launch a EUTELSAT-2 satellite in 1990.[20]

The Soviets also bid to launch for Inmarsat, the London based international maritime satellite organization, in which it is the fifth largest shareholder. There, the price offered for launch was $24 million. The United States officially objected to consideration of Proton use by INMARSAT. The contract was eventually awarded to McDonnell Douglas.

Most satellite owners have so far not been overly eager to take advantage of the attractive Soviet launch terms. Manufacturers are hesitant about turning over their technology to the Soviets for fear of compromise.[21] Regardless of customer interest or lack of interest, however, transport of any U.S. spacecraft—and foreign spacecrafts built under U.S. licenses—is forbidden under current U.S. export regulations, something which the U.S. government has stressed repeatedly and emphatically.[22] In September 1987, however, General Electric, a major U.S. satellite owner and manufacturer, urged the U.S. government to reconsider its policy through a written statement to the space science and applications subcommittee in the House of Representatives. Since in September 1988 the U.S. Department of Commerce approved an export license for a Hughes satellite to be launched on a Chinese Long March vehicle, this policy will come under even more intense scrutiny.

The Soviets have responded to technology transfer concerns with statements such as that from Nikolai Ryzhkov, chairman of the USSR Council of Ministers that "Western commercial payloads using Proton or other Soviet launch vehicles could be shipped in sealed containers to Soviet launch sites and exempted from customs inspections.[23] Soviet Official Dmitry Poletayev assured security of payloads during launch preparation as well as customer presence at Baikonur "up to the moment the rocket leaves the launch pad."[24] The Soviets have also stated that clients can test and store satellites in their own cleanroom, and may be 100 percent accompanied by Western technicians.[25]

Western manufacturers and government officials point out that even with technology transfer safeguards, significant technology must be exchanged before a satellite and a booster can be mated. Further, it is so far unclear as to whether foreign representatives would be allowed to be present during the mating process.

Since no contracts for commercial launches have been signed yet, excluding consideration of the Indian satellite, many launch terms are still unclear. It was not until very recently, October 1987, that the Soviets opened their launch sites to potential customers on an inspection-of-facilities basis, much like the Chinese have done with their Xichang launch site for some time.

Because of the problems encountered with marketing satellite launches, Glavkosmos began shifting their marketing emphasis to microgravity payloads in August 1988. This policy shift was intended to take advantage of the MIR station as a permanently manned space station and to employ automatic recoverable capsules for clients wanting to study microgravity processes. In February 1988, a contract was signed between the Soviets and an American company, Payload Systems Inc., to fly U.S. microgravity experiment payloads to MIR. Two recent offers from the Soviet Union illustrate the extent of Soviet commercial intentions. In April 1989, the Soviets demonstrated an increasing adeptness at typically Western commercial techniques by offering a "space advertising" package. For $620,000, Glavkosmos offered companies an advertising package which includes a three minute commercial filmed by the cosmonauts, patches on the cosmonauts' space suits, launch site billboards, and two 6 by 9 foot signs on MIR. Although out of character for the Soviets, the idea is evidently not without merit, as executives of a Japanese broadcasting firm immediately showed interest. They have, in fact, purchased a ride on MIR for one of their journalists—much to the chagrin of Soviet journalists who have yet to fly.

Also, in May 1989, the Soviets suggested that the U.S. Space Station could be launched on the Soviet Energia heavy-lift booster. Use of Energia, according to Soviet officials, would save NASA several billion dollars that they would otherwise spend on development of an unmanned Shuttle-C. Both of these offers evidence the Soviet desire to show space as an area where money can be made, as well as spent.

Potential Benefits in Opening the Soviet Program

The general Soviet position has been that no explicit changes have been made in their policy toward international cooperation in space. Commercial launches, it is said, were not offered in the past simply because internal launch demand made this prohibitive. Regarding space science, the Soviets say they have always been eager for cooperation with other countries.

In reality, even if the Soviets were not officially prohibiting commercial launches in the past, conditions which would have been imposed as part of the launch agreement would have made any offer moot. Now, however, the Soviets have said they are prepared to allow customers access to launch facilities. So the launch terms which the Soviets now say they are prepared to offer are certainly a change from the past.

Regarding cooperative space science projects, in the past the Soviets have worked primarily with Eastern bloc countries. Western scientists were, for the most part, not interested in cooperative projects anyway, here too because of the restrictive prohibitions under

which they would have had to work. The openness experienced by Western scientists and journalists during the Halley's Comet mission hopefully will be a precedent for the future. But the Soviets are not generally inclined to make policy changes without the potential for pragmatic benefits. These benefits are clearly evident regarding space.

There were two basic reasons that the Soviets closed their program to the West. First, the Soviet program has been, and continues to be, heavily dominated by the military. Until they deliberately allocated, and separated as much as possible, a limited part of the program for civilian activities, interaction with the West was politically impossible.

The second reason is that basic Soviet conservativeness made them extremely cautious about exposing themselves and their possible faults to others. This attitude was particularly exasperated by the success of the United States with the Apollo program.[26] Recognizing problems with their heavyweight launch vehicle which would have inevitably led to the United States reaching the Moon first resulted in the Soviets redirecting their efforts away from the Moon and into close-to-the-Earth, long-duration manned flights, where they have been very successful. But the spectacular achievements of Apollo created serious psychological hesitancy in the Soviets about exposing themselves to further potentially embarrassing situations, especially in an area where they had originally been viewed as the worldwide leader.

Now, however, the Soviets have justifiably achieved a high degree of confidence in their program. Their technology may be old, but it is proven. With more than 100 launches per year, the reliability of the technology justifies confidence in their program. The Soviets are now allowing live telecasts of launches, something unheard of in the past. So by separating out those portions of their space program that they did not consider necessarily under the domain of the military, and with their increased confidence in themselves, they are willing to break back into a field where the United States has been the leader, but is now experiencing difficulty. This is certainly not an insignificant political and public relations coup.

Because of their heightened self-confidence and by designating limited areas of a nonmilitary nature appropriate for cooperation, the Soviets are motivated toward joining the international space community by the potential for political, economic, and even psychological benefits. Involvement of Western scientists/countries with Soviet space science projects is a psychological boost for the Soviets, as it again, ostensibly, is Western acknowledgment that Soviet technology is as good as—or better—than that in the West. This recognition of parity has always been a factor in the Soviet psyche.

Politically, the Soviet Union was in the past its own worst enemy and inadvertently the United States' best friend regarding space. With the Soviet space program closed to the West, the United States had the exclusive capability to foster political and economic ties abroad through cooperative space ventures. One of the key reasons that the United States was able to hold a position of dominance for so long was that it had no serious competitor.

Now, Glavkosmos will actively pursue contacts, be they scientific or commercial, with developed countries and third world countries. Interaction with developed countries yields exposure to technology. Interaction with third world countries brings desirable linkages of a primarily political nature. In both cases there is also the element of prestige, much like the "positive image" purpose that has long motivated the United States.

Last, the Soviets have changed their attitude about space for a very simple reason: money. Cooperation and commercial launching at least partially relieves their own cost-intensive space budget. Further, space is an area where money can be made, and they would not be adverse to generating some hard currency. This has become especially important because of the loss they have experienced due to the drop in oil prices. The Soviets derive about 60 percent of their hard currency income from oil exports. A 1986 study reported that the Soviets would lose about $7 billion, or a third of its hard currency earnings, due to the collapse of crude oil prices that year.[27] Therefore, generating some hard currency through commercial launches could make up for at least a portion of this loss. The importance of raising revenue from space is also increasing as a means of justifying large expenditures. Some Westerners who have worked with the Soviets on space projects have in fact questioned whether "cooperation" is an accurate description of what is being sought by the Soviets, or whether it is really a polite way of saying "paid participation."

Implications of Change

Former NASA Administrator James Beggs said in 1985:

> They (Soviets) will be seeking to attract the international parties who have traditionally worked with us into an orbit with them to allow them to initiate activities with our European friends and even the Japanese in pursuit of (Soviet) goals in space.
>
> We will need to pay ever more attention to the issue of maintaining our competitive posture and continuing to maintain a lead so our friends around the world will want to work with us as their prime partner.[28]

The *Challenger* accident in 1986, and the current lack of direction of U.S. space policy, has unfortunately allowed Mr. Beggs's words to go unheeded as well as ensuring that his words are even more meaningful today.[29] But the Soviets are pressing ahead, anxious to step in with offers for cooperative space science projects where the United States would have in the past.

Although French–Soviet space cooperation has long been extensive, the scope of cooperation may be expanding to include ESA projects as well. There have been regular meetings between the U.S.S.R. and ESA over the years to exchange information on space science plans. In January 1987, agreement was reached concerning plans to include an ESA radiation dosimeter experiment on a Soviet Cosmos 1987 biosatellite.[30] Further, CNES (the French space agency) has held discussions with the Soviets to ensure the ESA backed Hermes spaceplane will have docking compatibility with MIR-type crafts, as well as the U.S./international Space Station. French cosmonaut Jean-Loup Chretien at one time even proposed that the Soviet Proton be used to launch Hermes, since the Ariane 5 booster will not be ready until the mid-1990s, and Hermes was to be operational before then.[31] Although unlikely, even serious consideration of that proposal would be a deviation from traditional patterns of cooperation.

Participation of international crews in Soviet space missions and manning the MIR station is increasing. Syrian Mohammed Fares participated in a joint research program aboard MIR in 1987, and Afghan Abdulla Ahad Mohmand traveled to MIR in 1988. Besides France, Bulgaria also has a cosmonaut currently training in the Cosmonaut's Training Center in the town of Zvyozodny, and discussions have been underway with Austria.[32] Certainly the strangest case to date of a potential cosmonaut is American singer John Denver. In 1988 Denver heavily lobbied the U.S. State Department in an effort to be allowed to spend a week at the Soviet MIR station.

. . . The More Things are Different, the More They Stay the Same

The Soviet penchant for not acknowledging problems has not totally abated, although in the period between 1986–1989 the Soviets showed signs of becoming significantly more candid about difficulties. The Soviets experienced two serious accidents in January 1987, which they tried to keep secret. On January 29, 1987, the Soviets deliberately exploded a military satellite to prevent the malfunctioning vehicle from falling into U.S. hands. On January 30, a Proton failure during the launch of a communications satellite was referred to as "the largest space vehicle accident since . . . Challenger and a USAF Titan."[33] They wanted to keep that estimated $50 million accident quiet to avoid hindering efforts to market the Proton as an international launch vehicle.

Another Proton booster failure on April 24, 1987, resulted in a disagreement of sorts between Glavkosmos officials and the Space Commerce Corporation. Glavkosmos denied the failure,[34] leaving Art Dula both to explain the technical failure to potential Proton customers initially with little or no information from his Soviet "partners," and to deal with the heightened image of the Soviets as secretive and perhaps untrustworthy business partners, an image he had already been trying to dispel.

On the other hand in September 1988, in what Western analysts considered a dramatic change, the Soviets were very open about the problems they encountered while deorbiting a Soyuz-TM spacecraft with a crew of two cosmonauts. According to *Aviation Week & Space Technology*, "During the tense hours when it appeared the cosmonauts were in danger, the Soviets disclosed their status and updated plans for the mission several times."[35] Apparently, just as with *glasnost* in other areas, change has started, but it will take time before becoming the norm—if it ever does.

Internal Soviet Space Politics and Problems

July 1988 was an important month for the Soviet space program. As part of a plan for the exploration of Mars, the Soviets launched two unmanned spacecrafts from the Baikonur Cosmodrome headed for the Martian moon Phobos, as the first step in an exploration process to culminate with a manned mission to the planet by the year 2010; the manned mission had also been announced in July 1988. The ambitious nature of the Soviet space agenda raised some eyebrows within the Soviet Union, where the domestic economic situation has reached a crisis magnitude. Yet the space program has also been a source of great pride for the Soviets; space activity traditionally has had a large following, which has grown even more because of the international admiration the Soviet space program recently has drawn. The Phobos spacecrafts carried experiments from eleven Western nations, including France, West Germany, and Austria, and several from Soviet bloc countries.[36]

Unfortunately, the Phobos I spacecraft was lost in September 1988 due to an incorrect ground control command sent to the spacecraft, which caused the Soviets to lose contact with the craft. In December 1988 some of the channels of the Phobos II craft's television system failed. That raised fears concerning the second spacecraft also. Moscow breathed a sigh of relief when nothing drastic occurred at that time, only to completely lose contact with the spacecraft later in March. The second spacecraft did relay a significant amount of data from its near circular orbit around Phobos before contact was lost. Preparations were in the final stage for the Phobos spacecraft to make a slow flyby of the moon, at which time landers would be released to perform a series of experiments. The loss of both spacecraft struck a substantial blow to Soviet plans for Mars exploration and prompted a reevaluation of their space program generally. In August 1989, the Soviets dropped the idea of a manned Mars mission around the year 2015, because of the technical difficulty and associated cost, and adopted a more conservative plan for planetary exploration than that which had been discussed before the Phobos loss.[37] Critics of the Soviet space program will continue to cite the Phobos losses to try and scale back plans.

In April 1989, IKI announced plans to prepare MIR for unmanned flight until June, and returned the crew to Earth. This was done in order to undertake a repair mission to fix electrical power problems on the facility. Problems with the solar arrays, which also caused problems for the U.S. Skylab space station in 1973, had begun to seriously hamper the amount of power available on MIR. Cosmonauts Alexander Vittorenko and Alexander Serebrov reoccupied MIR in September 1989. The continuous occupation of MIR as a manned space station has also been a source of internal debate over the need and utility of the big-ticket Soviet space program. This debate, combined with the technical problems, will undoubtedly result in reorganization among and within the Soviet space-related organizations.

IKI, directed by Sagdeev from 1973 until his retirement in 1988, and the Vernadsky Institute of Geochemistry and Analytical Chemistry, directed by Dr. Valery Barsukov, have been the two organizations traditionally in charge of the nonmilitary aspects of the Soviet space program, both under the umbrella of the Intercosmos Council. Even between those two organizations there were distinct differences, most evident by the stylistic and philosophical differences between the two individuals directing them. Whereas Sagdeev was, and continues to be, an active and vocal supporter of international cooperation,[38] particularly regarding space science projects, Barsukov has been just as clearly more nationalistic, with reservations about working with the West. Somewhat uncharacteristically, during a visit to Kennedy Space Center in May 1989, Barsukov did say that the Soviet Union hoped that the United States and other nations would join them in their planned Mars studies. The direction taken by the Soviet space program in 1985–86 reflects the confidence Gorbachev places in Sagdeev, who has also acted as Gorbachev's primary advisor on the U.S. Strategic Defense Initiative (SDI).

Glavkosmos now coordinates all work on civilian

space projects and is responsible for the operational use of space technology. Glavkosmos is said to be a complementary organ to the Intercosmos Council, which will still be responsible for the scientific aspects of space projects. Clearly, however, the establishment of another space organization adds another player to the Soviet decision-making circle and perhaps another competitor for leadership dominance. Evidence of an internal power struggle concerning space has been mounting. Although no longer at IKI, Sagdeev remains an active figure in the Soviet program and a vocal critic of aspects of it as well. At a November 1988 meeting of the Soviet Academy of Sciences, Sagdeev denounced the appointment of "figureheads" to major space-related positions, instead of experienced managers and scientists.[39] Glavkosmos is taking an aggressive role internally and has given indications that it may compete with Soyuzcarta, the Soviet organization which handles satellite images for commercial sales of Earth resource data.

At an April 1989 forum in Moscow, Sagdeev, who was intimately involved with the Phobos project from its inception, spoke about the poor coordination of project plans. The problem, according to Sagdeev, is that the Soviet military-industrial complex is more interested in satisfying hardware producers than making decisions about the space program based on merit.[40] This has led to large, secret budgets, which have drawn loud criticism and complaints from politicians such as former Moscow Communist Party boss Boris Yeltsin, and subsequently public opinion has turned against the space program. Within the Soviet Union, critics are voicing a new version of the guns-or-butter argument: that before the Phobos loss they were told that they would have bread once they got to Mars, but now it appears that they will have neither bread nor Mars.

Sagdeev and others have also spoken out against Buran, the Soviet space shuttle. Launched for the first time in November 1988, Sagdeev has denounced it as a waste of money, seemingly built at least partly on a keeping-up-with-the-Jones premise. Even Igor Volk, the shuttle's chief test pilot, has said that no clear purpose or use has been established for the shuttle as yet. Further, according to Volk, there are a substantial number of technical problems which remain to be worked out on Buran, more than initially evident by the success of its first test flight.[41]

Ironically, the Soviet space program is increasingly experiencing the types of political, economic, and public relations problems which have plagued the U.S. program for years.

Conclusions

Whereas other space programs have primarily matured technologically in this third period of examination, the Soviet space program has evolved politically, involving all the negative as well as positive ramifications that go along with that process. The Soviets are learning to deal with other countries in space relations, while simultaneously fighting major domestic turf battles. For the first time, the notion that space activity should be cost-effective seems to have crept into the Soviet planning process, accompanied by all the political scrutiny that entails. Clearly, reform and reorganization are being resisted in this field just as they are in many others.

The Soviet Union will have a much easier task of initiating cooperative space science agreements with Western countries than it will breaking into the commercial launch market. The technology transfers problems involved with commercial launches will make Proton commercial use the exception rather than the rule for some time. That the Soviets recognize this is evident in their shift of marketing emphasis to microgravity payloads in 1988. But science is another matter. Hungry Western space scientists who have seen many of their projects put on hold or cancelled altogether in recent years are eager for opportunities to fly their experiments. The notion of cost sharing is not unattractive either.

An increase in the number of actors in the international space community, with expanding capabilities among them, can lead to increased opportunities for cooperative space projects which otherwise might be cost prohibitive on an individual basis. Yet the political realities of protecting technology and linkage politics cannot be ignored. So whereas cooperation between Western allies presents an already difficult set of problems to be dealt with, potential cooperation with the Soviets adds an entirely new dimension to the process.

Endnotes

1. An earlier version of this chapter appeared in *Space Policy*, February 1988, pp. 60–73.
2. "Exploration of Outer Space for Peaceful Purposes," Chapter 29 in *Disarmament and Security 1988*, Institute of World Economy and International Relations, Soviet Academy of Sciences, 1988, p. 528.
3. Kenneth Pedersen, "New Challenges from Allies and Soviet to U.S. Space Supremacy," *Baltimore Sun*, 6 May 1987, p. 17A.
4. Walter A. McDougall, . . . *The Heavens and the Earth*, (New York: Basic Books, 1985), p. 273.
5. John D. H. Downing, "The U.S.S.R. and Satellite Communications," *Space Policy*, August 1985, p. 250.
6. ESRO/C/454, rev. 1. Adopted by the Council at its 34th session. Professor M. Keldysh's reply to Prof. H. Bondi's

letter is dated 12 March 1971. The agreement entered into force on the same date.

7. Downing, "The U.S.S.R. and Satellite Communications: Competition and Cooperation," p. 251.

8. Nicholas Daniloff, *The Kremlin and the Cosmos* (New York: Alfred A. Knopf, 1972), p. 187.

9. "Soviets Sign Intelsat Pact," *The New York Times*, 28 August 1985, IV, 12:1.

10. It is as yet unclear how events within the British Space Program, including the resignation of Roy Gibson as the head of the British National Space Center over budget concerns, will affect cooperative projects between the U.K. and the U.S.S.R. See: Roger Highfield, "What is Britain's Place in Space?" *Daily Telegraph*, 6 August 1987.

11. Specifically, a Soviet Pulsar X-1 spectrometer, the TTM imaging telescope from the U.K.'s Birmingham University and the Netherlands' Space Research Utrecht, the Sirene-2 gas scintillation proportional spectrometer from ESA's ESTEC technical center, and the Phoswich scintillation spectrometer, HEXE, from West Germany's Max Planck Institute for Extraterrestrial Physics and Tubingen University Astronomical Institute. Jeffrey Lenorovitz "Soviets Expected to Consolidate Space Program Activities in 1987," *Aviation Week & Space Technology*, 12 January 1987, pp. 97, 101.

12. "Soviets Offer Mission to British Astronaut," *AWST*, 16 June 1986, p. 22.

13. "British Astronaut to Fly On Soviet Space Mission," *AWST*, 3 July 1989, p. 17; "British Astronaut," *AWST*, 17 July 1989, p. 11.

14. An overview of past U.S.-Soviet space science cooperation is given in a statement by Mr. Richard J. H. Barnes, Director of International Relations, NASA, before the Subcommittee on International Scientific Cooperation, Committee on Science and Technology, U.S. House of Representatives, 25 June 1987, as well as a review of the provisions of the U.S.–Soviet governmental space agreement signed 15 April 1987.

15. "Soviets Profit by Using U.S. Space Technology," *AWST*, 6 October 1986, p. 23.

16. Phillip S. Clark, "China's Long March Into Space," *Space Log*, TRW, Vol. 22, 1986, p. 33.

17. Space Commerce Corp. announced plans in August 1989 to jointly develop and market (with a newly created Soviet organization, Technopribor) a new commercial mobile launch vehicle based on technology from the Soviet SS-20 medium range nuclear missile. See: John D. Morrocco, "U.S., Soviet Firms to Develop, Market Commercial Booster Based on SS-20," *AWST*, 7 August 1989, p. 21.

18. "Proton Marketing Team Finds U.S. Interest, Opposition," *AWST*, 25 May 1987, p. 20.

19. "Eutelsat Weighs Candidates for Alternate Launcher," *AWST*, 12 January 1987, p. 101.

20. "New Players Join the Launch Game," *Air & Cosmos Monthly*, July/August 1987, p. 81.

21. A group of former British Special Forces members have formed a firm—called Janus Inc.—to guard payloads at Soviet launch sites.

22. See, for example: *Interavia: Air Letter*, "American Niet to Soviet Launches of U.S. Satellites," (Geneva) 8 July 1987, No. 11,283, pp. 1–2.

23. "Soviets Discuss Proton Payload Exemptions," *AWST*, 12 January 1987, p. 33.

24. ". . . and the Soviets enter commercial launch market," *Aerospace America*, October 1986, p. 1.

25. See: "New Players Join the Launch Game," p. 82.

26. Nicholas Daniloff examines the Soviet attitudes toward space in his book *The Kremlin and the Cosmos* (New York: Alfred A. Knopf, 1972), particularly Chapters 5 and 6.

27. Lee A. Daniels, "Soviet Oil Export Rise Predicted by Consultant," *The New York Times*, 19 August 1986, IV, 1:1.

28. "Beggs Believes U.S.S.R. May Disrupt Western Bloc Space Unity." *AWST*, 4 November 1985, p. 18.

29. Reimar Lust, "The Cooperation of Europe and the U.S. in Space," The Fulbright 40th Anniversary Lecture, 6 April 1987, Washington, D.C., p. 5.

30. NASA will be participating on that mission also.

31. *Aerospace America*, May 1986, p. 1; "CNES Discusses Hermes-MIR compatibility with Soviets," *AWST*, 11 May 1987, p. 25. Now, technical development difficulties with Hermes is expected to push the first launch date until 1996, approximately two years after Ariane 5.

32. "Exploration of Outer Space for Peaceful Purposes," Chapter 29 in *Disarmament and Security 1988*, Institute of World Economy and International Relations, Soviet Academy of Sciences, 1988, p. 528.

33. Craig Covault, "Soviet Proton Booster Fails; Reconnaissance Satellite Explodes," *AWST*, 9 February 1987, p. 26.

34. "Glavkosmos Denies Proton Failure," *AWST*, 11 May 1987, p. 34.

35. "Soyuz Crew Lands Safely After Aborting Reentry Twice," *AWST*, 12 September 1988, p. 27.

36. *Aerospace America*, December 1986, p. 7.

37. Michael A. Dornheim, "Latest Soviet Planetary Mission Plans Reflect Shift to Conservative Outlook," *AWST*, 28 August 1989, pp. 21, 22.

38. Sagdeev, for example, is involved with the newly created International Foundation for the Survival and Development of Humanity, an organization that recommends cooperative solutions to global problems. See: Nicholas Daniloff, "The Space Statesman," *Air & Space*, November 1988, pp. 42–46.

39. *AWST*, 14 November 1988, p. 27.

40. J. Martin Sieff, "Space Funds Shrinking for Soviets," *Washington Times*, 28 April 1989.

41. Kathy Sawyer, "Soviet Shuttle's Mission Undefined," *Washington Post*, 30 April 1989, p. 1.

Chapter 11

Up and Coming Programs

In the high profile and high cost exploration of Halley's Comet, the space agencies from the "big four" space powers were represented: the United States, Soviet Union, Europe collectively, and Japan. Yet these are not the only countries which are active and increasing their space activities, and the Halley's group itself illustrates that "the players" in space are changing, with different motivations and capabilities. Clearly there has been an evolution from an environment twenty years ago where the United States and the Soviet Union were the sole players, to one where first Europe and then Japan joined the ranks of active participants. And the list does not stop there. China, Canada, Australia and an increasing number of third world countries are committing themselves both rhetorically and with government funds—the ultimate show of support—to expanding their space activities.

Canada

Canada presents an interesting example of a country which has had definite goals for its space activities, charted a course to meet those goals, and subsequently has been able to follow through on many of them. Canada is by no means a new player in space, rather it is one which has matured in recent years, according to plans laid out in the 1960s. Unfortunately, recent events in Canada also illustrate how even the best laid plans can be detrimentally affected by domestic politics.

Canada's geography has played a significant role in its space history. Its vast land mass has presented the Canadian government with major challenges with regard to transportation and communications. Early on, Canadians recognized that "space" could provide solutions to some of these problems and designed their space program to respond to those needs while simultaneously building a domestic space industry to yield financial returns. The latter part of that equation is particularly important because, like the United States and Soviet Union, financial resources for space activities in Canada are limited. Unlike the United States and the Soviet Union, however, Canada—like many other nations—does not augment its space-related budget through the military. Therefore, with limited financial and human resources, Canadians chose to focus their activities on specific space technologies to avoid spreading their resources too thin and hence render themselves ineffective in terms of progressing on the learning curve. They also chose to work cooperatively to share expenses. Canada's initial field of focus concerning space was telecommunications; later, robotics and automatic servicing would become an area of emphasis and expertise.

With regard to cooperation, Canadians point with pride to the fact that they do not have a single space project that is not in some way cooperative with another country. This, they say, has been a large factor in their success, as they can pick and choose what they want to utilize and participate in from services and projects offered by many countries. The Canadian space project most familiar to the general public is the remote manipulator arm built for use on the U.S. Shuttle, dubbed CANADARM. That project is part of an ongoing history of international cooperation between the United States and Canada, dating back to the Alouette and ISIS series of satellites. Development of the remote manipulator arm was funded by the Canadian government, under the direction of an industrial team led by Spar Aerospace Limited.[1] The Canadian government was willing to make such an investment to build the learning curve of Canadian scientists, engineers, and industry in robotics and artificial intelligence—a field not only relevant to space, but with almost unlimited potential on Earth as well. Canadians have been consistent in their development of a course of action whereby space activities have a direct positive feedback domestically.

Extending their experience with robotics and servicing via the remote manipulator arm, Canadians will now contribute a mobile servicing system (MSS) to the Space Station It includes a remote manipulator and mobile base, a special purpose dexterous manipulator, external and internal astronaut workstations, in-orbit systems for power management and distribution, data management and communications, a ground operations center, and an operations management and logistics center which will include a major simulation facility. "The importance of this program in terms of the impact it will have well into the next century in broadly based space operations, technology development and technology 'pull' in knowledge intensive fields is well recognized by the Canadian space community."[2] Canada clearly states its objectives of participation as including the advancement of Canadian capabilities in strategic technologies associated with the MSS, production of a state-of-the-art system, and development and exploitation of the associated economic spin-offs.[3]

The importance and future potential for robotics experience were also recognized in the United States during the Space Station negotiations. The chairman and the ranking minority member of the House subcommittee responsible for NASA funding went so far as to send Acting NASA Administrator William Graham a letter stating "deep concern that U.S. industry be able to make maximum use of the space station as a vehicle for vital spinoff benefits in the automation and robotics field."[4] Obviously, the final agreement between the United States and Canada had to reflect the concerns and desires of both sides. In that regard, NASA will develop the Flight Telerobotic Servicer (FTS) as part of the Space Station program. Although U.S. and Canadian facilities are being designed to perform different tasks, there will be some overlapping capabilities.

The Canadian space program is perhaps most noteworthy and commendable in that Canada is one of the few countries in the world where the space sector brings in more economic returns from abroad than the government spends on it. For 1985, for example, private space-related industries brought in about $320 million in sales, 70 percent of which was generated from exports. During the same year, government expenditures relating to space were approximately $150 million.[5] A communications satellite for mobile users, called MSAT,[6] has attracted over $400 million in private capital. Canada's space industries, over 100 of them, have been thriving. By far the most successful has been Spar Aerospace, Ltd., the largest Canadian space company, which has a contract for a $125 million Brazilian satellite and another for ESA's Olympus satellite worth $65 million.[7] With respect to returns from investment, the Canadian space program presents itself as a model for other countries to follow.

Unfortunately, but perhaps not surprisingly, after building a space program centered around pragmatism and careful planning, the Canadian space program has

recently shown itself as susceptible to domestic political disputes as any other. Canada has long stated an intent to create a space agency. Until recently, jurisdiction for space activities was scattered throughout twelve different departments and agencies within the Canadian bureaucracy, with an interdepartmental committee on space established in 1969 as the coordinating mechanism. Plans for a Canadian space agency had been bantered about since 1969, but it was not until Prime Minister Brian Mulroney's Throne Speech in October 1986 that a space agency was targeted as a way to heighten his government's profile in science policy, and hence seemed as though it might finally become a reality. Almost immediately, however, the as-yet-unborn space agency became the focus of a power struggle concerning where it should be located. Ottawa and Montreal both actively lobbied to be chosen as the site location, ostensibly because of the jobs that the agency would bring (although a high estimate was about 300, the low estimate was 35), but in reality the motivation was primarily prestige.

Montreal business and political leaders rallied in a huge public relations campaign where "Montreal, C'est Spatial" (Montreal Is Spatial) was the theme. National political parties became involved in the issue also, voicing their preference. The tug-of-war between the cities actually held up the legislation creating the agency. Christopher Trump, then Corporate Affairs vice-president at Spar was quoted as saying "Where to put the agency is the government's business, and we won't intrude. But we wish something would happen soon."[8] The space agency finally came into being early in 1989. Against the advice of many people involved in the Canadian space program, Montreal was selected as the site for the agency headquarters. This places it right in the middle of the French-English language controversy which currently dominates the Quebec and Canadian political scene.

Dr. Larkin Kerwin, president of the National Research Council of Canada since 1980, has been chosen to head the new agency. Much of the task of ensuring that the transition from the old system to the new agency runs smoothly, and that Quebec politics do not affect the fledgling agency, will fall to him. One of the potential benefits of having a space agency is that it can act as a unified advocate for space activities. It is hoped that will be the case for this heretofore model space program.

Pacific Basin

The countries of the world perhaps drawing the most attention because of their ambitious space-related plans lie in the Pacific Basin: China and Japan. It is expected that both countries will be concentrating their efforts over the next five to ten years toward becoming major international space powers. Already active, when the space programs of these countries mature to the degree that the European program has, there will be as much a change in the patterns of cooperation and competition in space as when the Soviets opened their doors to the West in 1985–86 and when Europe obtained independent launch capability with the Ariane.

Japan

Japan has two space agencies, the National Space Development Agency (NASDA), a government financed corporation for research and development which operates under the Science and Technology Agency of the Prime Minister's Office, and the Institute of Space and Aeronautical Science (ISAS), a research institute operated through the Ministry of Education. Both agencies work closely with the Ministry of International Trade and Industry (MITI), which has a separate Space Industry Division, and all space activities are coordinated by the Space Activities Commission within the Prime Minister's Office.

The principles which the Space Activities Commission has set provide the foundation of the Japanese space efforts. They are 1) that Japanese space efforts will be solely for peaceful purposes, and in fact they have no budget for military space activities, 2) that Japan will develop indigenous and autonomous space capabilities, and 3) that space science and technology is to be developed in harmony with international society and international cooperation. Hence their current efforts are focusing on satellites, development of the indigenous H-II launcher,[9] the Japanese contribution to the Space Station, JEM, and defining an aerospace plane, possibly in collaboration with other countries.

According to a 1987 long-term policy report conducted under the auspices of the Space Activities Commission, a primary goal of Japan's space policy is the development and utilization of outer space. That challenge, according to the report,

> will contribute to the harmonization of Japan's industrial structure with international economy and thus will realize a more comfortable and affluent society of Japan, through the innovative progress of science and technology and the improvement of land use.[10]

Japan says it seeks to increase "the size of the pie" rather than competing for already existing shares, concerning space activities. Space is seen as a potentially lucrative business, one that should be expanded. Japanese philosophy is that space should be promoted from a user friendly and responsive perspective, rather than rely on government needs and funds as most countries, particularly the United States, have done in the past. The Japanese space budget is approximately

.04 percent of their GNP, as compared to approximately 2 percent of the U.S. GNP. Therefore the size and amount of increase of the Japanese budget spent on space is small in comparison to U.S. and European space budgets. But private sector funds spent on space is estimated at approximately $23 billion, and is increasing in both size and proportion.

In conjunction with those figures and their anticipated growth, MITI has undertaken policies to promote Japanese space industries, specifically development, utilization, and manufacturing. Japanese companies are anxious to expand their space activities, including work on manned space activities, as evidenced by Mitsubishi Heavy Industries recently setting up a 9-meter deep tank at its Kobe shipyard for simulated space walks. MITI's plan is to coordinate government plans with private industry early on, with the goal of making Japanese space industries nonreliant on government funds by the mid 1990s. To do that:

> The Government will, with financial prospect, promote development deliberately so that private enterprises may be able to conduct precursor activities with the prospect for the future. It will also apply various measures such as financial investments and loans, advantageous taxation system, the completion of large-scale testing facilities for governmental and private common use, transfer of technology, and various technological information service to encourage technological development, space utilization activity, etc., conducted by private enterprises.[11]

The reason for this careful posturing now is simple. The Japanese estimate the scale of world space industry in the year 2000 to be at Y1,000 billion annually.[12] Clearly, the Japanese intend to be prepared to reap their share.

It does need to be mentioned, however, that changes in the makeup of the Japanese government since the 1987 long-term policy report was issued have resulted in hints of budget "constraints" that were not evident prior. Although Japan's Diet gave final authorization during the summer of 1989 for the $2 billion budgeted for the Japanese experiment module (JEM) to be an attached lab on the Space Station,[13] it is not certain that if Space Station negotiations were being held in 1989, the attitude of the Japanese government would have been as enthusiastic as it was in 1986–88. Certainly MITI will continue to push the government to keep with the accelerated pace of R&D outlined for the space fields, but it remains to be seen how the government will respond.

China

China launched its first satellite on an indigenous rocket in 1970. Since that time the Chinese have made steady progress both in launch vehicle design and other areas of space technology. They were the third nation, after the United States and France, to fly a cryogenic hydrogen/oxygen upper stage. The vehicle was the Long March 3 rocket, first launched in 1984. With that, they have also placed two communications satellites in geostationary orbit. Currently, China is working on development of a Long March 4L rocket, comparable to an Ariane-4, targeted for a 1991 launch.

China's activity in space is notable in its bid to be an active competitor in the international commercial space market. In September 1984, China successfully flew and retrieved a French experiment on a Chinese polar orbiting satellite, launched on a Long March 2. This was the first commercial low-orbit microgravity payload ever flown. China has since carried various international materials processing payloads to orbit, including a (French) Matra microgravity test platform and a (West German) Intospace protein crystal growth experiment. China also plans to launch a Mailstar data relay satellite for Sweden.

In January 1989, a Hong Kong company signed a $30 million contract to launch an American-built satellite into space, called AsiaSat 1, on a Long March 3. The launch, which took place in April 1990, was the first Chinese launch of a foreign-built satellite. China will also be launching two other American-made satellites into space for an Australian satellite company, AUSSAT, in July 1991 and February 1992. AsiaSat 1 and the two AUSSAT satellites were made by the U.S. based Hughes Aircraft company.

In a controversial 1988 decision, the United States approved the necessary export licenses for U.S.-built satellites to be launched on the Chinese boosters. This was done after Hughes proposed a plan to protect its satellites from Chinese access during the launch process. Under the plan, China will forgo customs inspections of the satellites and Chinese technicians will not be allowed to touch or have access to the satellites.

The area of the most concern to American private industry, however, is the feeling that the Chinese are undercutting them with inexpensive, promotional pricing and insurance rates for launches. In that regard, meetings were held in 1988 concurrent with those to discuss the export licenses, discussing the placement of quotas on Long March space boosters that can be sold to launch U.S.-built satellites. The agreement that was reached allows for China to be able to launch up to nine international communications satellites through 1994. Although this will limit the impact China can have on the overall market, it also acknowledges and establishes China as an active competitor.

Over the next ten to twenty years, China will expand its space program with the advent of new launchers, including the CZ-2E, which will be capable of placing 19,000 pounds into low Earth orbit. This is more than enough to launch a manned spacecraft. Work being done at Chinese space medicine institutes support efforts toward a manned program, and it is expected that Chinese astronauts could be selected within the next five years.

The Chinese government says that it intends to hold to its plans for development of its aerospace industry and will continue to contract for foreign satellite launches post-Tiananmen Square, but Western reaction remains uncertain. On June 5, 1989, President Bush temporarily suspended the export license allowing Hughes Aircraft to ship the first of its three satellites to China for launch.[14] The annual meeting of the International Astronautical Federation (IAF) which had been scheduled to be held in Beijing in 1989, was relocated in Malaga, Spain in reaction to the Beijing crackdown.

China clearly sees space as an integral part of its overall development strategy. The government is actively supporting the promotion of the aerospace field, in ways such as increasing annual budget allocations and setting up special funds for key areas of research. For this to be done, despite the nation's financial austerity, space must truly be seen as critical to its future.

Australia

In the 1960s, Australia was one of the world's leading nations in space research. Not only was it the third country to launch a scientific satellite from its own territory, Australian scientists were also deeply involved with experimental projects. But when the United States and ELDO stopped using the Woomera launch site in the early 1970s, Australian politicians seemed quickly to lose interest in funding further space activities.

The mid-1980s saw a resurgence of interest in space research in Australia, prompted primarily by scientists. Australia discussed forming a space agency, intended to consolidate national efforts and use space as a kind of "sunrise industry." Typically, however, the problem has been matching good intentions with adequate funding. Because of that situation, a rather innovative approach to increasing space activity has come about in Australia.

Since ELDO first intended to launch their EUROPA rocket from Woomera, Australians have realized the potential of their country as a launch site. In 1988 the Australian Spaceport Group, made up of Australia's largest company, the Broken Hill Proprietary Company Limited (BHP), with Bond Corporation, Comalco Limited, and the U.S. aerospace giant Martin

Marietta Corporation, formed to carry out comprehensive investigations concerning building a spaceport on Cape York. Cape York is in the state of Queensland, in Australia's northeastern sector. The unique aspect of this whole venture is that it is not the federal government in Australia that is leading this effort to try and get into the space business, but the State of Queensland. The approach taken by Queensland has been to offer legislation to enable a spaceport to be built and to finance support structures, such as roads, but to have the facility owned, operated, managed, and developed by private enterprise rather than by direct investment from the state itself. Clearly, space as an elite venture open to only a very few countries is no longer the case.

Third World

The allure of space for developing countries has become increasingly apparent over the past several years. Clearly, their interests focus on the "practical" aspects of space, including: telecommunications for business, education, medicine, and other purposes; territorial and weather remote sensing; and launcher services. Quite simply, space is seen as a technology driver, a way to improve the domestic quality of life, and a symbol of national prestige. For some countries, the allure is also in the military aspects of space.

In May 1987 Israel test-fired a medium-range missile into the Mediterranean Sea. Although initial reports focused on the potential of the missile to carry a nuclear warhead far enough to strike Soviet territory, it seems the test had another purpose also. Israel was testing the country's ability to launch a surveillance satellite into low earth orbit. Now, in 1989, Israel is on the verge of becoming the eighth country with the capability to put a satellite into space. The missile, called "Shavit" (Hebrew for "Comet") is built by Rafael, Israel's leading missile manufacturer, and Israel Aircraft Industries. Israel's motivations are clear. Having the ability to monitor Arab activity would give Israel a distinct military advantage over its neighbors. Further, this capability would free Israel from dependence on U.S. satellite intelligence. Although Israel has had a space agency since 1983, its activities are almost exclusively focused on this satellite project.

India, with its long-standing space program, leads the developing countries in space, a position it has worked hard for and covets. India has built two different launch vehicles, plus its own remote sensing satellite, and is now working on its own communication satellite. The program stresses Earth applications: communications, remote sensing, national resource survey and management, and meteorological monitoring. The fact that India, with a population of over 750 million, roughly 75 percent of which lives in rural areas, having extremely high illiteracy rates, and a per capita income of only approximately $170 annually, commits relatively vast amounts of money to space activities evidences its belief that space will play a vital role in its future. Using space for socioeconomic improvements has been repeatedly stated as the underlying rationale since the inception of this space program. Other countries are beginning to follow suit.

Brazil has the most active space program in Latin America, with activities dating back to the 1960s. EMBRATEL, the Brazilian communications satellite organization operated through the Ministry of Communications, operates two satellites built specifically for Brazil by Spar Aerospace in Canada. The Brazilian Ministry of Aeronautics has facilities for the launch of sounding rockets, often used in conjunction with work conducted by the Brazilian Institute for Space Research. In 1967 Brazil began operation of its first reception station for data from meteorological satellites. In 1987 that station was adapted to receive data from the French SPOT system also.

Between approximately 1963 and 1973 civilian space efforts in Brazil were very much comparable to those in India. However, the Brazilian civil space program was generally allowed to stagnate between 1973 and 1983, when the government was under military rule. Although in 1979 the Brazilian Complete Space Mission (MECB) was approved by the government, only since around 1985 has the government been able to get its space program back on track.

The MECB was originally conceived as a joint program with France. Rather quickly though, it was decided that the timetable and subsequent resources which Brazil would have had to commit to keep up with French plans were more than what was feasible. Therefore, in what is described as a "friendly split"[15] Brazil decided to proceed on its own at a slower rate. The goals of MECB include: four experimental satellites for data collection and remote sensing; an independent launch vehicle; and a launch base. Recently, Brazil has started looking for other third world partners for cooperative space ventures. China and Brazil, for example, are currently working together on a $150 million Earth resources satellite.

Whether the desire is there or not, many third world countries simply do not have the financial or human resources to develop independent space programs on a broad scale. The United Nations Committee on the Peaceful Uses of Outer Space (COPOUS) has provided somewhat of an outlet for developing countries in space affairs, but primarily from a legal perspective. What is needed, however, is a way for scientific and technical capabilities to be increased in the third world through more active participation. Consortia have been repeatedly suggested as a mechanism to accomplish this goal. ARABSAT is an example of how in the 1980s countries with relatively small space programs began to work together regionally regarding communications satellites. In 1985 several African nations joined together to study how space could benefit their region. Their focus has been primarily on communications and resource management through remote sensing. Also, a Pan American Space Organization (PASO) has been proposed, along the lines of ESA, to strengthen participation of Latin America in space.

Conclusion

There appears to be a consensus among many of the industrialized, or industrializing, countries that space holds many of the technologies which will drive economic growth in the twenty-first century. Hence, in preparation, these countries are positioning themselves to increase their capabilities, whether in particular areas or over a wide range of areas. The environment is ripe for these countries to begin or to intensify their activities because options regarding other countries to work with, to progress on the learning curve, are expanding.

The benefits that countries foresee are basically the same as those which pushed Europe into cooperative ventures in the 1960s: economic, political, and technological—separately and in combination. In Japan and Canada there have been concerted efforts to privatize the space effort, in the sense of making it increasingly less dependent on government funding, and in fact a money-making venture. These countries will provide models for others following behind them. It makes far more sense to try and emulate countries which have profited from space financially as well as technologically, than those where space has been viewed as a budget drain.

Within the developing nations, there has been and will continue to be a primary emphasis on using satellites for communications and land use planning. The third world nations will be a major market for both space hardware and services. Expansion of their participation in space ventures outside these areas will most likely be through consortia, or regional groups, to consolidate both funds and manpower.

Endnotes

1. The agreement was that NASA was then to procure production systems for the Shuttle fleet from Canada.
2. Ronald W. Neville, "Canada in Space," Presented at Joint AAS/Japanese Rocket Society (JRS) Symposium, 15–19 December 1985, Honolulu, Hawaii, p. 13.
3. F.R. Vigneron, et al. "Technology Research and Development Associated with the Mobile Servicing System," Fourth Canadian Aeronautics and Space Institute Conference on Astronautics, 3–4 November 1987, Ottawa, Ontario, p. 1.
4. Letter from Edward Boland and Bill Green to William Graham, 14 January 1986, p. 2.
5. R. Bryan Erb, "Space Business in Canada," Speech presented at the Houston Space Business Roundtable, 16 December 1987, p. 4.
6. The MSAT project became the focus of a dispute when the U.S. Federal Communications Commission ruled that satellite mobile services in the United States could no longer use the 800 Mhz band, which Canada strongly preferred for MSAT, but were instead directed to part of the L band. That meant Canadian MSAT services could not be utilized in the United States. Appeals were made to the FCC and talks were held at Spar Aerospace about building dual-band satellites. All-in-all, however, the cooperative spirit at the Canadian Department of Communications had seen better days.
7. Bill Gladstone, "The Canadian Space Program Takes Off," *McLean's*, 8 April 1985, p. 49.
8. Bruce Wallace, "An Intercity Space Squabble," *McClean's*, 4 January 1988, p. 10.
9. In conjunction with the H-II development, a massive construction project is underway at the Tanegashima launch site to build a suitable launch pad complex. The complex has been compared to the U.S. Air Force's facility at Cape Canaveral, Florida, used for the Titan launch vehicle.
10. *Toward a New Era of Space Science and Technology,* (unofficial translation) Report of the Consultative Committee on Long-Term Policy Under the Space Activities Commission, 26 May 1987, p. 6.
11. Ibid., pp. 22–23.
12. Ibid., p. 18.
13. James Harford, "NASA Reassures Edgy Space Station Partners," *Aerospace America*, September 1989, p. 11.
14. K.K. Chadha, "China Holds Course for Space Program," *Aerospace America*, September 1989, p. 7.
15. "Brazil's Space Program Remains Dynamic Despite Fiscal Woes," *Aviation Week & Space Technology*, 24 August 1987, p. 77.

Chapter 12
New Commercial Challenges

A New Kind of Competition

In the 1950s and 1960s, "competition in space" basically had to do with the "space race." The United States and the Soviet Union were in an undeclared war to outshine each other in space achievements—culminating for the United States with a trip to the Moon. This "race for prestige" gave the requisite initial impetus to get space activity going, but provided little support for further space activities once detente replaced cold war politics. Although prestige is still an important factor—and often a driver—in space activity, "competition" in relation to space activities has now taken on a new meaning, one far more related to the commercial aspects of mature space programs with advanced capabilities.

A fact of life in the present environment in which cooperation on space ventures takes place is that competition has also become a reality to be dealt with. As most of the competition is commercially driven, it is important to review where the emphasis is, and where it might be in the future. Whether or not policy makers will be able to accept cooperative and competitive activities taking place in peaceful coexistence is one of the most important questions yet to be answered.

Post-Challenger Accident Commercial Attitude

The truth of the statement that "The lack of an assured source of transportation to and from space is the greatest roadblock to space manufacturing"[1] was vividly demonstrated by the impact of the *Challenger* accident on plans for commercial space activity. The commercial development of space basically became a dormant issue during the first part of the Shuttle hiatus. It became hard to keep up appearances of activity and enthusiasm when the U.S. space program was grounded. The *Challenger* tragedy clearly pointed out the vulnerability of space transportation and the dependency of commercialization, as it was envisioned in 1984, on transportation.

After the initial shock of the catastrophe and the realization of the vulnerability of the U.S. space program wore off, the basic consensus from the business community and policy analysts was that commercialization prospects were not as bright as had been thought before, but were not as bleak as initially thought afterward either. There was a shift in emphasis, post-*Challenger*, which was probably one that needed to be made anyway. People began to accept, as had been repeatedly pointed out by space business analysts like John Egan, that there is no such thing as "space business"; rather there are businesses which will use space.

> A pharmaceutical company interested in using space to separate a drug is not in the space business; it is simply using space as a place to do work. The company that is involved in producing semiconductor materials in space is using space to improve the semiconductor that it is trying to sell commercially. Commercial business is using the space environment.... What is good for the semiconductor industry using space may not be good for the medical and pharmaceutical industry using space. Success in one area in no way ensures success in any other.[2]

From this realization came the ability to divide those organizations and companies involved in commercialization into two groups: users and services. Whereas prior the emphasis had been almost solely on encouraging and assisting users, who were dependent on access to space for their ventures, post-*Challenger* emphasis began to shift to services. Services include ground and support operations, as well as planning for long-term ventures such as space construction. Concurrently, the access problem in both the long and short term can be dealt with so that users will be able to go into space to do things like crystal growth.

Space Transportation

After the *Challenger* accident and the return to use of expendable launch vehicles (ELVs) and the entrance of other countries into the launch market, the space transportation policy question changed from whether there was sufficient launch capability to service the commercial satellite industry, to whether or not the demand would be sufficient to keep the market alive. The size of payload capacity necessary to be competitive on the commercial launch market is changing. Most requests for proposals (RFPs) five years ago were for satellites weighing 1,100–1,400 kg. During this time, competition was primarily between the Delta, Atlas Centaur, Shuttle, and the Ariane 1 to 3 launchers, which could deal with those weights very adequately. Increasingly over the next five years, however, satellites in the 2,000–2,800 kg range will dominate the market, due mainly to the increase in power and numbers of transponders.[3] Launchers must then be able to lift a 2,000 kg payload to geostationary transfer orbit (GTO) to be competitive for the larger communication satellites. It is expected that eventually advances in technology will decrease the weight of satellites, making sales for launchers like the Delta and Long March 3 at least partially dependent on whether satellites are designed within their lower weight range. The market for commercial launches currently is about twenty satellites per year, of which about 80 percent of those are to GTO. The total market is made up of approximately 70 percent telecommunications satellites (GTO), 20 percent Earth observation satellites (about half of which go into GTO), and 10 percent research satellite into various orbits. By 1995 the GTO market is expected to be about twenty-three commercial launches per year. This then raises questions as to whether there will be an adequate market for candidate launchers. According to at least one analysis, a launcher family needs approximately twelve satellites annually (which corresponds to about eight launchers) to achieve a sound commercial return.[4] Clearly then there will be competition within this field. The situation will become even more complex as more launchers become commercially operational.

Several general factors should be kept in mind when making comparisons between launchers. For example, payload launch mass capacity is affected by geographical latitude of the launching place. Kourou, French Guiana (about 7 degrees N) is better than Cape Canaveral (about 29 degrees N) for the same launches to geostationary orbit because you get a greater velocity from the Earth's rotation when launching from sites nearer to the equator, thus reducing the amount of fuel needed to obtain orbital velocity. This factor is evidenced in the large disparity between lift capacity of the Soviet heavy lift Proton to geostationary equatorial orbit and GTO. The difference is largely due to the additional energy required to achieve GTO from the latitudes of Soviet launch sites.

The issue of government subsidies is always a source of concern when considering the "level playing field" question involved with international launch competition. Although subsidies certainly cannot be ignored

Table 12.1 Currently Operational Commercial Launchers

Type	Approximate Payload Capacity to GTO	Approximate Cost	Approximate Reliability
Proton (Soviet) 4 stage	2200 kg* 4600 kg	$24–40 M	92% since 1970
Long March 3 (China)	1400 kg	$35 M & up	67% of proven flights
Ariane 3 (Arianespace)	2600 kg (dual 2400)	$40 million & up**	Ariane 1/2/3 79% over 14 operational flights
Ariane 4 (Arianespace)	4200 kg (or 1369 kg + 2270 kg for dual)	$80 million plus for dual launch***	first launch 6/88
Titan III (Martin Marietta)	1900 kg* 4500 kg	$90–115 M	96% over 135 launches
Atlas-G (Gen Dynamics)	2180 kg	$59 M	90% since 1957
Delta II (McDonnell Douglas)	1447 kg	$50 M	93% since 1960

  * to geostationary equatorial orbit
 ** (½ for double launch = 250 Mil FF; full = 460 Mil FF)
*** Plans announced in January 1989 by Arianespace to order 50 Ariane 4 launchers would make the cost of a launch at between $90–100 million.

Table 12.2 Future Commercial Launchers

Type	Approximate Payload Capacity to GTO	Status
Titan IV (Martin Marietta)	8200 kg	Currently operational; military use only
Ariane 5 (Arianespace)	6800 dual: 5900	Available 1994 or after
H-2 (Japan)	8000 kg to LEO 3864 to GTO	First test flight planned for 1992
Atlas G/Centaur (General Dynamics)	2358	Initial commercial launch planned summer 1990

as a reality for the United States to contend with, it is highly unlikely that it would be seriously considered as an option in the United States. That increases the importance of other factors which can make U.S. ELV use attractive. Also, some manufacturers, such as Martin Marietta and McDonnell Douglas, are able to at least partially underwrite their commercial ELV expenses with military launch and vehicle production and refurbishing contracts, which some people say is a kind of indirect subsidy.

The time a customer must wait to get on a launch manifest is also a consideration for commercial launches. The wait can very from at best 12 months for a Long March and 20 months for a Proton, to 36

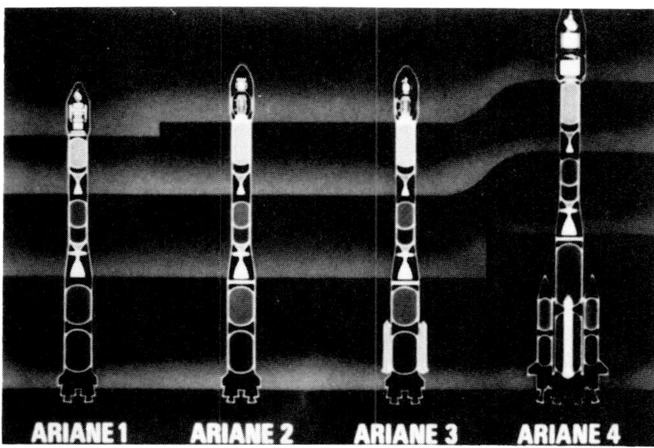

Figure 12.1 Ariane I–IV family of launchers. *Photo courtesy of European Space Agency.*

months or more for an Ariane or most U.S. ELVs. Customer manifest wait is not only a function of how long it would take to have an available launcher and integrate the technology involved. Launch site availability is also a factor. On that basis, General Dynamics, for example, has decided to invest $18 million in a second launch pad at Cape Canaveral. This would increase its launch capability from three to five launches per year to eight to ten per year.[5]

Another consideration when comparing launch vehicles is insurance. Even before the *Challenger* disaster, "satellite malfunctions and launch failures leading to multimillion dollar payouts [were] raising cern that the market [would] not be able to react to demand."[6] At that time, launch insurance rates ranged from 17 percent to 20 percent of the launch price, if insurance could be raised at all. Currently, insurance rates can go as high as 35 percent on the open market. The Soviets and the Chinese are both offering lower priced, more attractive insurance arrangements than the West. The Soviets, who consider launch and spacecraft insurance separately, have said a premium of 12–15 percent of the launch price would "guarantee a free reflight or launch cost return, (minus the insurance premium) in the event of a booster failure."[7] Europeans have access to special insurance rates through Arianespace. This puts U.S. launch vehicles and users at a monetary disadvantage. Clearly, the competition within the space transportation field is becoming increasingly complex and intense and will continue to do so.

Examination of ELV competition is, however, looking at the present; we need to look to the future. The access to space problem does not just involve the reliability rates of ELVs; it really involves the need for cheaper, better, more efficient, and diverse access to space. In 1988, it cost approximately $4000 per pound to put a payload in low Earth orbit. That figure makes going into space to work on anything that could otherwise be done on the ground almost prohibitive. Until a figure of about $200 per pound to low Earth orbit can be reached, space will remain an exotic workplace economically out of reach of most entities. If the cost per pound to orbit figure is brought down, however, some experts believe the possibilities are endless. "When people start talking in terms of getting costs down to hundreds of dollars per pound, then you'll see the largest industrial revolution that has ever been around," says Dr. David Webb, former member of the National Commission on Space and president of the International Hypersonic Research Institute. "You'll see everybody going to space."[8]

There are a variety of designs being suggested for next generation space flight, beyond a next generation shuttle or a more advanced Ariane V, with hypersonic flight gaining increasing support. Hypersonic vehicles are basically horizontal takeoff vehicles with the ability to travel at speeds over Mach 5. These, however, could only be high speed terrestrial transports. The space vehicles must be capable of Mach 25 speeds to reach orbit. Several countries have designs somewhere between the conceptual and experimental phases.

The United States has the X-30, or national aerospace plane,[9] (dubbed the Orient Express by the Reagan administration for public relations purposes) as its hypersonic technology program. In 1989, the Department of Defense tried to kill the $300 million FY 1990 budget request for the X-30. Behind the scenes maneuvering and quick intervention by the National Space Council reversed the DOD decision. More critical budget decisions are yet to come, however, because when the X-30 moves from the conceptual and planning phase into the actual building of a vehicle, budget requests will significantly increase. The decision will be made in 1993 as to whether or not to go ahead and build an X-30 experimental technology-demonstration vehicle. DOD feels that the project is too commercially oriented and futuristic, and subsequently gives other projects with nearer term application potential a higher priority. Others, however, stressing the importance of hypersonic technologies for the next century, insist that any hesitation with this program will put the United States behind other countries active in the field, including Japan, Europe, and the Soviet Union.

The Germans are working on the Sanger two-stage design,[10] British Aerospace is trying to gather an international consortium to support their unmanned HOTOL version, and the Japanese and Soviets are involved in hypersonic research also.[11] Soviet efforts appear to focus on a Tupolov design for a Mach 5–6 transport, although recently interest in hypersonic aircraft for surveillance and interception was expressed by Soviet Marshall of Aviation Alexander Yefimov.[12] Soviet officials visiting Japan in August 1989 to discuss possible cooperative space ventures included joint development of an aerospace plane on their agenda. Although it is unlikely that such a partnership would be made, Soviet interest certainly illustrates the interest and importance with which they view hypersonic technology.

Hypersonic vehicles all involve cutting-edge technology and are still in the technology development phase. The nation which best takes advantage of that planning period to lead the way now will have the competitive edge in the future.

The Future

Commercial space activity has continued since the *Challenger* accident and is in fact growing throughout much of the rest of the world. In the United States, however, what growth is taking place seems to be in spite of, rather than because of, the U.S. government. All four of the major space powers, the United States, Japan, ESA, and the Soviet Union, are currently positioning themselves to be competitive in whatever long-term markets emerge concerning space commerce. The problem is, ESA and Japan have significant government-industry partnerships and space is entirely a state enterprise in the Soviet Union; this as opposed to the situation in the United States where the government makes policy and private industry is left to deal with it.

Former NASA Administrator James Fletcher summarized the situation:

> In order to fully appreciate the commercial situation, it is essential to understand the institutional differences which distinguish NASA from most of its foreign partners. NASA is committed to stimulating and promoting commercial development of the U.S. space industry. However, . . . many foreign space agencies stimulate and promote their industries through direct intervention in the planning and decision-making processes. Such intervention gives them certain latitudes not available to NASA.[13]

Most space agencies work for their federal Department of Industry or some similar entity. Joint private-public entities such as COMSAT, Arianespace, and SPOT Image are becoming increasingly utilized elsewhere as models for successful commercial space ventures. But in the United States it has only been since 1984 that NASA has had an Office of Commercial Programs, and perhaps more importantly, only in November 1988 that a Commercial Programs Advisory Committee—basically an industrial advisory committee—was formed. The success of Comsat shows that the United States can cope with public-private institutions if it has the political will. But unless the United States adapts a more flexible attitude, with regard to both its

institutions and ways of doing business, it will clearly be at a disadvantage in future space commercial ventures.

Japan, noted for its ability to maximize the "follower" strategy of business over the last decade, has realized that the follower strategy will not work in space and has taken steps to become a leader in that field. One aspect of leadership in space has come to mean having a competitive advantage. Although the United States has seemingly recognized this, as evidenced by President Reagan's 1988 National Space Policy, it has been incredibly slow to do anything about it.

Officials in and outside of the government have recognized the need to redefine the relationship between the U.S. government and the private sector concerning space. Robert H. Brumley, former general counsel of the U.S. Department of Commerce and chairman of the Commercial Space Working Group, has succinctly stated what he sees as the proper relationship.

> There is a need for the government's infrastructure to encourage companies to go into space or at least to allow one to create the stepping stones to get there . . .
>
> The government is a very bad business partner. It does not understand business plans. It is not motivated by the profit motive . . .
>
> The essence of commercial use of space or of the encouragement to invest in space is to continue to decide in what areas we need to get government out of that business plan, to get them away from the table and let companies compete at arm's length with those who also want to go to space.[14]

Actually doing something toward redefining the relationship has been more difficult.

The National Space Policy, approved on January 5, 1988, directed the continued and expanded development of a separate, nongovernment commercial space sector through private sector investment. It was greeted with much applause and enthusiasm, in the hope that space was finally going to move from the necessary, government-sponsored development stage to the also necessary privatization stage, much like the airline industry did fifty years earlier. In reality, the U.S. government has become the greatest impediment to the flourishing of the U.S. commercial space sector, both by its failure to follow through on its stated commercial policies and by the mixed signals it sends to industry as to what the government will do for industry and what it expects from them. Consequently, industry representatives say this undercuts confidence in the U.S. government commitment to space business, and companies interested in commercial space business might have problems raising outside capital to support their projects or be unwilling to commit their own resources to such projects.

Another aspect of the privatization issue involves

differing intragovernmental attitudes and degrees of support for privatization. Although Congress and the White House have directed NASA to keep its costs down on projects like the Space Shuttle and Space Station, and congressional funding for NASA is precarious at best, in 1989 Congress continued to balk at the idea of using private contractors to finance or lease portions of its proposed space station project, as it did in 1984 when NASA began discussing privatization of the Shuttle. Members of Congress say that NASA is only trying to play budget tricks under the guise of commercialization. NASA admits there is no long line of investors waiting to make proposals, but says that the White House has directed them to privatize projects that might otherwise require government funds. The result is that NASA often walks a fine line between White House and congressional directives.

Conclusion

The United States has long had a reputation as a nation involved with both research and development of new technologies and their subsequent products and market applications. Japan, on the other hand, has excelled at the latter part of that process, market applications, allowing others to open the field initially. But in space, Japan has decided that it cannot afford to follow past strategy. Japan wants to be a leader in space, as do many other nations. Why? Because the stakes are perceived as too high in the long term to accept anything less than a front runner's position. The United States, on the other hand, is clinging to old institutions and ways of doing business in a way totally out of context with traditional American innovativeness and spirit— at a time when it can least afford to do so. Clearly, the U.S. government and the private sector will need to establish a better working relationship if the United States is going to be competitive in the international aerospace field in the future.

The reality of the situation is that opportunity costs money; one invests money to make money. Somebody is going to make money in space in the future, if not within the United States, then somewhere else. Therefore, how much it costs to get the American civilian space program back into a leadership position, in terms of technology, flight opportunities, services, exploration, and so on, should not be an issue. Rather, the question under scrutiny should be how much it will cost in the future, in economic, political, and scientific terms, NOT to do something. The answer to that question is, in all respects, too much.

Commercial competition in space ventures is something that the United States has only recently had to become accustomed to. This has created some uncomfortable feelings which have spilled over as a sense of protectiveness in cooperative ventures. But protectiveness and commercial drive are not limited to the United States; all space-faring nations would like to see a return on their substantial investments. How this attitude will influence or possibly even jeopardize future cooperative ventures has clearly become another factor for consideration.

Endnotes

1. John E. Naugle, "A Manufacturer's View of Commercial Activity in Space," in *Economics and Technology in U.S. Space Policy*, Molly K. Macauley, ed. (Washington, D.C.: Resources for the Future & the National Academy of Engineering, 1987), p. 70.
2. John J. Egan, "Conducting Business and Scientific Experiments in Space," *Space: National Programs and International Cooperation* (Boulder, CO: Westview Press, 1989), p. 137.
3. See: Robb Stoddard, "Arianespace: On Center Stage," interview with Frederic d'Allest, *Satellite Communications*, June 1987, p. 20, and Arianespace, Launch Market Profile-Commercial Satellites 1987–1993, November 1986, DC1274, Figure 3.
4. *Reaching for the Skies*, European Space Agency, Paris Cedex 15, France, BR-42, 1988, p. 14.
5. "New Players Join Launch Game," *Air & Cosmos Monthly*, July/August 1987, p. 81.
6. Herbert Coleman, "Underwriters Tallied $600 Million in Unexpected Insurance Losses," *Commercial Space*, Fall 1985, p. 61.
7. *Aviation Week & Space Technology*, 25 May 1987, p. 21.
8. Quoted in "Pushing Onward With Space Travel," Richard Burnett, *Orlando Sentinel*, 20 March 1989, p. 10.
9. See: *National Aero-Space Plane: A Technology and Demonstration Program to Build the X-30*, U.S. General Accounting Office, April 1988, GAO/NSIAD-88-122.
10. The rationale for the two-stage Sanger is as a candidate for the next generation European launch vehicle. Hermann A. Strub, Deputy Director General, Head of Aerospace Programs, West German Federal Ministry for Research and Technology (BMFT), "Hypersonic Technology in the German Space Program" presentation at the International Aerospace Symposium, Nygoya 89, 14–16 February 1989, p. 7.
11. For information on these programs see: *Proceedings of the First International Conference on Hypersonic Flight in the 21st Century*, 20–23 September 1988, University of North Dakota, Grand Forks, N. Dak.
12. Richard DeMeis, "Aerospace *Glasnost*," *Aerospace America*, January 1989, p. 23.
13. Statement by James C. Fletcher, NASA administrator, before the Subcommittee on Space Science and Applications, Committee on Science, Space and Technology, U.S. House of Representatives, 10 December 1987, p. 5.
14. Robert H. Brumley, "Future Directions in Space Commerce," *Space: National Programs and International Cooperation*, Wayne C. Thompson & Steven W. Guerrier, eds. (Boulder, Col. Westview Press, 1989), pp. 156–157.

Chapter 13

Space Station Negotiations: The New Considerations of Cooperation

On January 25, 1988, in his State of the Union Address, President Ronald Reagan announced that the United States would build a space station, and invited "friends and allies to participate in the development and use of the space station." Although the admirable intent of President Reagan's invitation should not be overlooked, especially in light of the fact that it was against the advice of some of his advisors, the limitations of the invitation should also be understood.

First, there was no mandate given in the president's speech that the Space Station be international, only an invitation made. Second, "participate" does not necessarily equate with "partnership." And finally, friends and allies were invited to participate in the *development and use* of the Space Station, with no mention made of management. The implications of these limitations quickly became apparent when the negotiations for Space Station participation began.

To illustrate the problems involved with cooperation on space projects—even among allies—in the new international regime and on an open-ended project, the Space Station negotiations between NASA and ESA will be examined from President Reagan's invitation of participation through the signing of the intergovernmental agreement on September 29, 1988. Although Canada and Japan are also Space Station "partners," the details of each set of negotiations are unique, and the experiences of those other than ESA will not be included here except in a broad sense.

The initial phasing and scheduling for the Space Station was as follows: Phase A, January 1984–April 1985, to discuss general interest and basic concepts; Phase B, April 1985–April 1987, for design and definition, with contracts to be awarded; and Phase C/D, development, to begin in April 1987. The station was to have initial operating capability (IOC) early in the

Figure 13.1 Space Station and Shuttle. *Photo courtesy of NASA.*

1990s. The scheduling plan began to be postponed almost immediately, however, because of funding problems.

Early in the negotiations, philosophical differences between the United States and its potential partners became evident. NASA wanted to focus on program and technical negotiations, while ESA wanted to work on basic concepts of design purpose and principles, and then fit the hardware in later. Which approach one takes can make a big difference in the eventual design and utilization scheme of the station.

Originally, negotiations were opened under two fundamental premises set by the United States. The first was that, in order to maximize space and resources, there would be "functional allocation" of the station, meaning that facilities for the same purpose would not be duplicated by participants. The second was that all station elements would be open to all parties. Again, the idea was to get maximum use from the station.

The Past Catches Up With the United States

As it has already been stated, the ESA Council meeting at the ministerial level held in Rome, January 1985, adopted a long-term European space plan. The long-term program was intended to lead to "a comprehensive autonomous European capability in space and containing the following major elements: in-orbit infrastructure programme, space transportation systems programme and programmes for earth observation, telecommunications, microgravity, space science and technology."[1] It also included endorsement of the Columbus program as an optional ESA program, with the intent that Columbus would be the European contribution to the Space Station. This endorsement, particularly pushed by the Germans, was obtained as part of a package deal arrangement. The French agreed to support the Columbus program if the Germans would support development of a cryogenic engine for a new generation of Ariane launchers (Ariane 5) and the Hermes spaceplane, both primarily French programs.[2] Clearly, Europeans intended to work toward autonomy in space, but also recognized their own technical and financial limitations and the benefits of still working with the United States on the Space Station and other appropriate projects.

Although their overall enthusiasm for the Space Station project was evident from the start, the Europeans were also not at all subtle in making it clear that these negotiations were going to be different from those which preceded the Spacelab agreement.[3] Statements from Europeans like "We in Europe still remember the unsatisfactory deal we had on Spacelab, so we're being very careful not to rush into a new agreement until all aspects are clearly defined,"[4] repeatedly showed up in

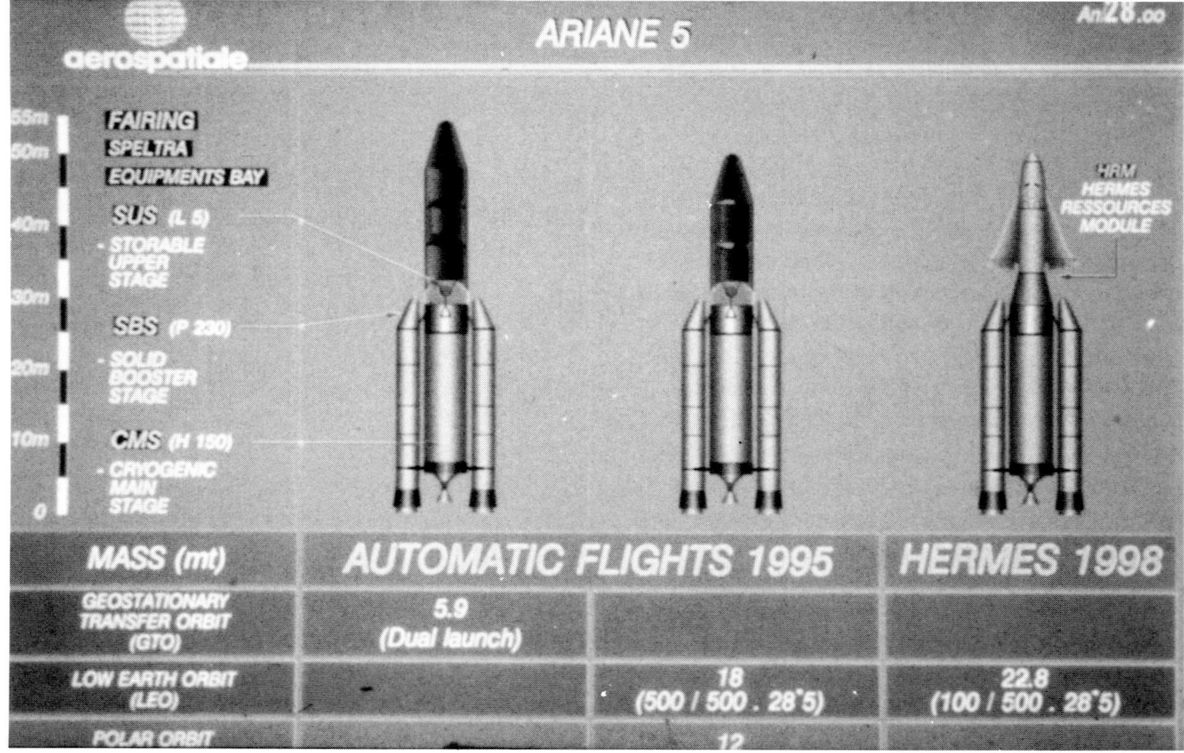

MASS (mt)	AUTOMATIC FLIGHTS 1995	HERMES 1998
GEOSTATIONARY TRANSFER ORBIT (GTO)	5.9 (Dual launch)	
LOW EARTH ORBIT (LEO)	18 (500 / 500 . 28°5)	22.8 (100 / 500 . 28°5)
POLAR ORBIT	12	

Figure 13.2 Ariane V with and without Hermes configuration. *Photo courtesy of European Space Agency.*

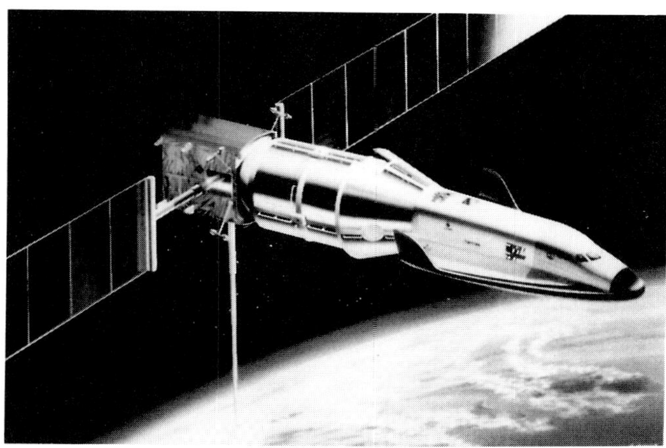

Figure 13.3 Hermes deploying satellite. *Photo courtesy of European Space Agency.*

news articles.[5] U.S. negotiators remembered the Spacelab agreement also, but primarily from a different perspective: as Europe being graciously allowed to participate in a U.S. manned spaceflight program, an area previously reserved to the United States and Soviet space programs.

The Europeans entered the Space Station negotiations with the following set of conditions for cooperation:[6]

- Appropriate participation of Europe in the Space Station Programme;
- European responsibility for design, development and operation of the ESA-provided elements;
- Access to all space station elements on a reciprocal, non-discriminatory basis;
- Satisfactory sharing of operations costs;
- Satisfactory arrangement for technology transfer;
- Reciprocal arrangements to off-set supplies and services between Europe and the U.S.A.;
- Achievement of maximum legal security and parity of governmental commitments.
- Guaranteed availability of U.S. Space Transportation system and communications facilities with the possibility of later using corresponding European facilities.

The fundamental philosophical principles behind these points were each traceable back to a prior experience with the United States.

Partnership

The Europeans made it clear to NASA during Phase A that they were interested in cooperating in the Space Station program as an integral partner in the Space Station initial operating capability (IOC). They wanted Europe to take over essential Space Station IOC elements and not just to provide additional/complemen-

tary Space Station elements or functions. To that end, ESA undertook an extensive definition study of its potential participation in the program. This desire can be traced back to its involvement with the Shuttle program, where it wanted to build an IOC but was deemed by NASA not to have the requisite experience or capability to do so. Because of its desire to be involved with the PAP ESA was willing to take on a clearly secondary part of the project, Spacelab. By 1985, however, the European program had clearly matured and now did have the experience and capability to play a more integral role in the Space Station.

The Europeans realized that NASA would probably not be amenable to the idea of their contributing an IOC, because of the technology transfer problems and simply because this was not how it was used to operating. How to define what would be acceptable for Europe as a "significant contribution" and whether or not to push the IOC aspect as a prerequisite for participation was a decision they knew they were going to have to make early on.

The Europeans basically saw three courses that their participation could take: providing a laboratory module that docked to the U.S. station and met European needs; providing a module which met joint U.S./European requirements; or taking responsibility for a key subsystem component of the station. In progressive order, those options range from the simplest interface arrangement to the most complicated. On that basis, the Europeans recognized that the first option would be the most attractive to the United States. That option eventually became most attractive to them also. Building a laboratory module would assure them more independence from the United States than subcontracting for actual space station components—independence to develop technology that they could later use on their own space station as part of their plans for an autonomous program.

European demands for a Space Station "partnership" were to a certain extent viewed as hypocritical by NASA and others. Dr. Wulf von Kries of the German Aerospace Research Establishment, writing about the lack of cooperative spirit on both sides of the Atlantic, pointed out:

> The Europeans have repeatedly stated that they want a "genuine partnership" with the U.S.A. Yet with their insistence on separate ownership, jurisdiction, use and control of the space station elements they intend to provide, it is quite obvious that a businesslike partnership is not on their minds. The relationship the Europeans see, in fact, seems more that of a privileged associate who, for only a limited contribution of use of his labour property, is granted a disproportionately large measure of directional rights and material benefits.[7]

Other Europeans, however, are quick to point out that

even if Europe—or Japan or Canada—had wanted to be an equal partner, that was not an option. NASA wanted control. Therefore, according to this line of thought, if their contribution was going to be limited and largely determined by NASA, the Europeans would have been remiss to do anything other than protect their own interests as much as possible. So although it often seemed as though the majority of the negotiations focused on legal issues, sometimes of a speculative and even nit-picking nature, more often than not the big picture, political issues were the dominant ones.

Right of Access

The Europeans wanted to make sure that they would have a guaranteed right of access to the entire Space Station, and that the access could be used for commercial purposes. After the Sparx venture, the Europeans were naturally concerned about how U.S. laws would affect their use of the Space Station.[8] As commercial potential was a dominant reason for their wanting to become involved with the station, their concern was understandable.

Management

Since this is an open-ended project where cooperation inherently extends beyond development, unlike Spacelab and most projects in the past, the Europeans wanted to make sure that they would have a voice in the management of the station over time. This would become perhaps the most difficult negotiation problem to deal with.

Commitment

Before allocating funds to the project, the Europeans wanted a guarantee from the United States that it was willing to commit funds to building, completing, and operating the Space Station, and that the commitment was irrevocable. This point was obviously in reference to their experience with the United States cancelling its spacecraft for the ISPM project. A European aerospace executive was quoted as saying, "We still aren't forgetting that NASA didn't live up to its commitments to Europe when it cancelled the U.S. spacecraft part of the ISPM several years ago due to funding constraints."[9] The fact is, however, that there is no absolute guarantee the United States can give to international partners because in the United States the allocation of funds for this or any project is done on an annual basis.

The Europeans, especially since ISPM, have recognized the dilemma of the U.S. budget process in their negotiating groups. Clearly, no international agreement, treaty, or executive agreement can commit the resources of the United States; this comes under the strict authority of Congress. That the Space Station would be a matter of significant debate in Congress seemed clear to everyone from the start. When some members of Congress began asking questions like, "How can we build houses in space if we cannot afford houses on Earth?" NASA, as well as ESA, Japan, and Canada, became understandably nervous. The Europeans saw as an ideal solution one involving a multi-year funding agreement between Congress and NASA. Recognizing that the U.S. Congress would be very reluctant to give up a major source of power, its budget prerogative,[10] other solutions were envisaged, including termination clauses and clauses for indemnification of expenses incurred should one party default. It is rather ironic that in the Spacelab agreement, concern focused on Europe as the party likely to default, whereas now concern focuses on the United States.

The Europeans concluded that to maximize commitment from the United States a treaty would seem the obvious choice of document to be used. The time it would take for Congress to act toward that goal would be prohibitive, however, and therefore they agreed that an executive agreement would have to be acceptable. It was noted by the Europeans, however, that if the executive agreement did not provide for intervention by Congress that the agreement would be no more binding that an MOU and hence offer little as a guarantee of commitment. In Europe the executive agreement, or Inter-governmental Agreement (IGA), will have the status of a treaty, whereas in the United States, the IGA cannot be upheld in the national courts.

Further, the Europeans were insistent that the external signs which reinforce an international agreement, in terms of the perceived political significance attached to it, were not to be overlooked either. The U.S. negotiators, certainly not unaware of European concerns, suggested that there were various ways of "visually enhancing" the value of an intergovernmental agreement, including the status of the signatory (president, vice-president, ambassador), the surrounding publicity, and the political support that can be given through "joint" or "concurrent" resolutions from Congress. All these mechanisms had been used in the past, sometimes more successfully than others. There was, however, little that the United States could offer as an alternative.

Two major documents were being negotiated between NASA and its potential partners—the IGAs and the MOUs for each phase. The IGAs were negotiated primarily by politicians and lawyers, Deputy Assistant Secretary of State (Bureau of Oceans and International Environmental and Scientific Affairs) Richard Smith chaired the IGA negotiations for the United States, which was appropriate because that department has more extensive experience in dealing with political and legal issues than does NASA. The MOUs, as implementing documents for the IGAs, were concerned with the more technical aspects of the project and were negotiated by the space agencies involved. Each document references the other where appropriate.

The United States began by trying to negotiate three separate IGAs rather than one multinational IGA for all partners. In October 1987 they finally agreed on one IGA, but with three separate MOUs for implementation. Most Europeans felt that the IGA was the key document and should be given primary attention in the negotiations. The U.S. position, however, was that the IGA was basically a showcase document or at most at the same level as the MOUs. ESA refused to yield this point, and now the IGA is considered the primary document, though symbiotic to the MOU.

The European Contribution

Although Europe had wanted to keep its options open until the end of Phase B, the Phase B MOU which entered into force 3 June 1985 limited the station elements open to European development. The Columbus Preparatory program was to consist of studies on a pressurized module, payload carriers in low orbit and polar orbit, and resource modules. Initial discussions between NASA and ESA centered on the laboratory module as being a free-flying or detachable version of Spacelab. NASA was not at all pleased with that idea, based on suspicions that since the Europeans wanted autonomy in space, ESA planned to use the U.S. Space Station temporarily and then, when capable, detach the module and leave the Space Station. OMB had basically the same attitude, that the United States would get burned, but with a new twist. OMB wanted NASA to entertain the idea of international participation in the core program first, to prevent any of the participants being able later to leave and second, to bring down the IOC costs. NASA rejected that idea, wanting to maintain complete control of the core configuration.

Eventually, it was agreed that ESA would contribute three Space Station flight elements: an attached pressurized module (APM), a polar platform, and a man tended free flyer (MTFF). The APM is to be permanently attached, while the MTFF will be capable of autonomous operational periods of six months or longer.

As Phase B got underway, the United States began to examine the implications of the Space Station more carefully from a legal standpoint, the major concern being jurisdiction.

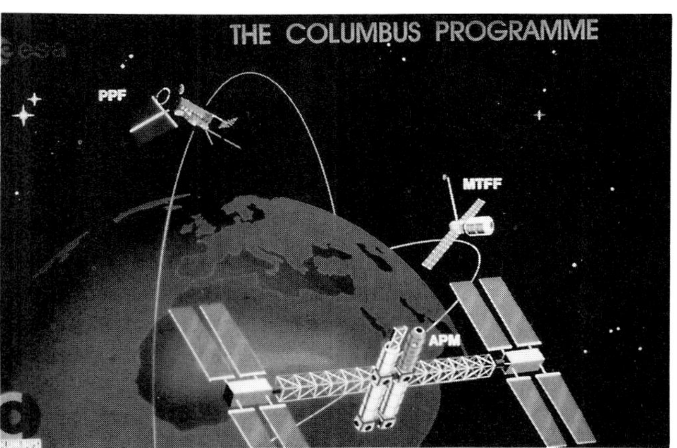

Figure 13.4 Columbus program elements. *Photo courtesy of European Space Agency.*

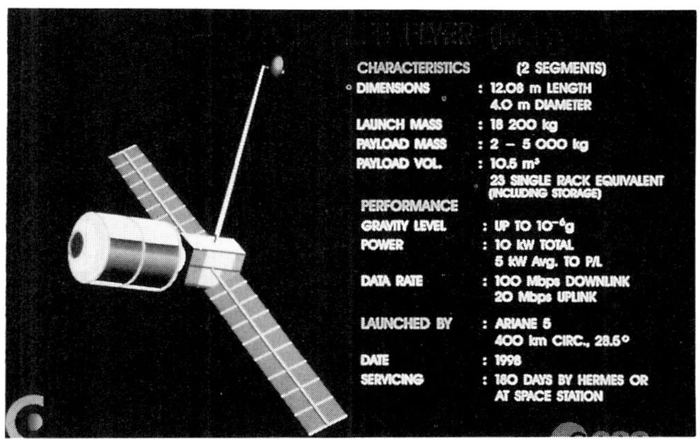

Figure 13.5 Man-tended free flyer. *Photo courtesy of European Space Agency.*

Jurisdiction Issues

At the request of the Senate Committee on Commerce, Science and Transportation, the congressional Office of Technology Assessment (OTA) prepared a background paper on "Space Stations and the Law: Selected Legal Issues."[11] This background paper dealt not only with jurisdiction issues, but commercial, criminal, and tort issues as well. OTA held a workshop in May 1986 to review the finding of its draft report. At that time, the workshop participants agreed that:

> Determining jurisdiction (i.e., deciding which nation has the right to make and enforce rules of law) is the single most important legal question to resolve in the planning stage for the first space station. Although a legal concept, jurisdiction with respect to an international space station will involve important—and sometimes overriding—technical and foreign policy considerations.[12]

The Europeans recognized that jurisdiction was of primary importance as well. To a large extent, the 1967 Outer Space Treaty was used as a foundation from

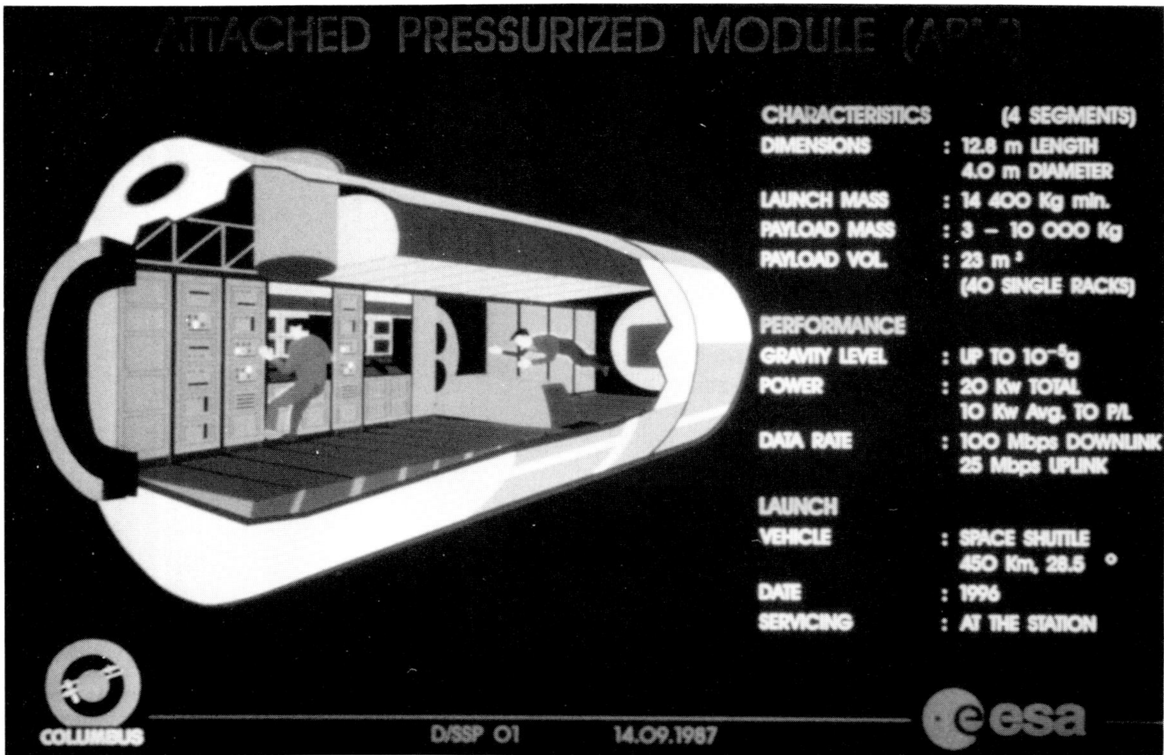

Figure 13.6 Attached pressurized module. *Photo courtesy of European Space Agency.*

which to draw legal principles concerning the Space Station for all the negotiators.

It was generally concluded that no proprietary rights and, accordingly, intellectual property rights could be asserted based on national law in outer space. Therefore the crucial points became an agreed definition of "space object" and "jurisdiction and control." Clearly, whoever legally registered a "space object," as the space station would certainly be defined, would have jurisdiction over that space object. So were the elements of the Space Station to be considered separately or as a single unit?

OTA concluded that there were at least four different types of legal status which a space station could have:

- a national space station under the jurisdiction and control of a single nation;
- a multinational space station under the joint jurisdiction and control of several nations;
- a multinational space station the individual modules of which are under the independent jurisdiction and control of separate nations; or
- an international space station under the jurisdiction and control of an international governmental organization similar to INTELSAT.

They went on to state that "To avoid controversy and complexity of cooperative international ownership and operation, the United States may wish to retain complete control over the space station."[13] The United States did indeed wish to do so.

Commercial Interest Issues

In the Spring 1985 issue of *Commercial Space*, Hans Hoffman, then managing director of the space division of the German industrial conglomerate MBB/ERNO, is quoted in reference to the Space Station as saying, "No one is under any illusions about how long it will take to earn a commercial return. But it must be done eventually. Nothing will be done for pure research."[14] Future economic return was clearly the condition upon which the West German government agreed to participate in the U.S. Space Station project and would be a precondition for support of future space projects as well. This notion was not unique to Germany either.

In light of past experiences with the United States, particularly regarding the SPARX project, Europeans were understandably concerned that they would have the right to commercially exploit discoveries made on the Space Station. Legally, the issue focused on which patent laws will apply to Space Station discoveries. The Europeans initially believed that applying U.S. patent laws to European research findings could make it impossible to utilize them commercially.[15]

Other Legal Issues

The range of legal issues raised during the Space Station negotiations was such that one European official remarked that perhaps the best contribution that ESA could make to the station was not a module for working on microgravity or life sciences, but a module for working out legal problems. The NASA Legal Office identified at least nine areas to be considered: jurisdiction and control, criminal law, civil causes of action, interparty waiver of liability, third party liability, insurance, proprietary rights, technology transfer, and taxation,[16] and encompassed within these subdivisions are many others. For example, civil causes of action include such areas as torts, workman's compensation, and domestic law—imagine the complications of a space-based paternity suit—for consideration. Dealing with the issues was particularly difficult because they were all based on hypothetical situations.

Phase C/D Negotiations

The Phase C/D negotiations can basically be divided into four periods: Summer 1986–December 1986; January 1987–May 1987, June 1987–November 1987; and December–January 1988.

Phase 1

The primary issues being dealt with during this period were concerned with reciprocal rights for utilization and participation and with operating costs. NASA supported the idea of functional allocation of facilities, which was strongly opposed by ESA. ESA's objections were: (1) that aposteriori cost accountability would be difficult; (2) that it would be difficult for it to "sell" Space Station participation to its Member States if they did not know how much ESA could use the facilities; (3) that ESA intended to maintain primary control of its own elements; and (4) defining operations costs would be difficult, e.g., if ESA was to pay 20 percent, the question became 20 percent of what?

By the end of Phase 1, functional allocation was dead.[17] It was basically then agreed that each contributor would use its own labs, but that NASA would get some use of all labs. Further, on the question of operations costs responsibility, NASA and ESA agreed to share responsibility based on a percentage of common system operating costs, that is, those attributed to the operation of the manned base as a whole. Negotiations were, for the most part, going well. At that point, however, DOD actively entered the picture, accusing NASA of being too easy on ESA in its concessions.

Phase 2

In December 1985, the U.S. Defense Department suddenly demanded explicit assurances that military research on the Space Station would not be prohibited.[18]

Although unable to state exactly what type of research they wanted to be able to do on the station, after the *Challenger* accident and the subsequent grounding of the Shuttles, DOD clearly wanted to keep its options open. Secretary of Defense Casper Weinberger sent a letter to Secretary of State George Schultz concerning the Space Station. The letter, which was widely disseminated,[19] talked about national security concerns and expressed the view that the United States would pay too high a price for international cooperation under certain conditions. These conditions included:

* Fail(ing) explicitly to reserve the right to conduct national security activities on the U.S. elements of the Space Station, without the approval or review of other nations;
* Accede to multilateral decision-making on matters of Space Station management, utilization, or operation;
* Permit a one-way flow of U.S. space technology to participating nations who are also our competitors in space; or
* Allow the concept of "equal partnership" to displace either the reality or symbol of U.S. leadership in the Space Station program.[20]

Consequently, in February 1987, use of the Space Station for "national security purposes" was included as part of the NASA negotiating proposal. Beyond the immediate problems this "advisory letter" created in the Space Station negotiations, the letter also exemplified the growing chasm between NASA and DOD and the growing influence of DOD in civil space matters.

It was during Phase 2 that the Space Station was cut back from the original dual-keel design to one of "phased development." This, subsequently, meant that the ESA ratio had to be cut back also. When the decisions were made in that regard, ESA was informed, but not involved. Critical to ESA during this period was that NASA accepted the idea of ESA contributing a man tended free flyer (MTFF). The MTFF had originally been rejected as competition to the industrial space facility (ISF) being considered for use by the United States. ESA considered it essential and it became a bottom line prerequisite; on that basis NASA agreed to allow it as part of the European contribution.

Phase 3

Between June and November 1987 no headway was made on negotiations, and in fact the negotiations came perilously close to breaking down. The block in the negotiations concerned management decision making for the development, utilization, and operation of the station elements. During this phase, the United States, through NASA, became more aggressive. It took the stance that if ESA wanted to contribute three elements, the pressurized module, polar platform, and MTFF, all must be treated the same as far as management. ESA felt that it should have the final say in decisions concerning its elements in development, utilization, and operation. NASA felt that as the primary station contributor, it should have the final word on all decisions, including those concerning development.

The ESA delegates met at The Hague in November 1987 and said no to that U.S. position. In fact, they got confirmation at that time to go forward with the Columbus program with or without the United States.

Phase 4

In Phase 4, a compromise was reached regarding the management of the ESA contributions to Space Station and regarding operations costs. The compromise was that ESA would yield to the United States regarding the final decision on development, operations, and utilization matters concerning the attached pressurized module, and ESA would have the final management decision on the MTFF and polar platform, except where servicing was concerned.

Management was dealt with in Article 8 of the MOU.

> 8.1.a. NASA and ESA each have responsibilities regarding the management of their respective operations and utilization activities and the overall Space Station operations and utilization activities, in accordance with this MOU. NASA will have the responsibility for the overall planning for and direction of the operation of the manned base . . . and the NASA provided Polar Platform. ESA will have the responsibility for the planning for and direction of the operation of the elements it provides which are separated from the manned base {specifically, the MTFF when outside the operational CCZ (Command and Control Zone) of the manned base and the ESA-provided Polar Platform when outside the operational CCZ of the STS . . . }
>
> 8.1.b. A Multilateral Coordination Board (MCB) will be established as soon as possible after the start of NASA's Phase C/D and will meet periodically over the lifetime of the program or promptly at the request of any partner with the task to ensure coordination of the activities of the partners related to the operation and utilization of the Space Station.

Concerning operations costs, it was agreed that the categories of operations/services to be included in the calculation of operations costs are included in the MOU (Article 9), and no categories can be added without the agreement of both parties. Further, each contributor pays 100 percent of the operations and maintenance costs for its own facilities.

Provisions were also included so that parties could purchase from or barter with other parties for use of their facilities. ESA was satisfied, however, in that it was to begin from a "preferred position," meaning that

it would know what it was starting with, in terms of allocation time for each piece of hardware contributed: 51 percent of the attached pressurized module, 100 percent of the MTFF and polar platform. NASA could purchase up to 25 percent of the MTFF time, and use of the polar platform was to be on a balanced reciprocal basis with other platforms.

Finally, a compromise was also reached regarding the use of the Space Station by DOD. Agreement was made that the Space Station is a civilian project to be used for "peaceful purposes." "Peaceful," however, is a noticeably ambiguous term. Therefore, it was further agreed that the nature of individual projects to be undertaken on the station would be determined before they commenced, with the final determination to be made—whether or not something is a "peaceful" project—by the party in whose section the research or activity would take place.

Conclusion

On September 29, 1988, intergovernmental agreements between NASA, ESA, Japan, and Canada were signed in Washington, D.C., making them partners in the use of the Space Station for the next thirty years. Secretary of State George Schultz signed for the United States. The timing was less than optimal from the standpoint of publicity, in that it was on the same day as the launch of *Discovery*, returning America into space after a two and one half year absence. Lack of publicity and public understanding of the role of the Space Station in the larger scheme of space exploration and development have always been a contributing factor to the program's more fundamental larger problems: commitment and hence funding. Although all the partners survived the negotiations, it now remains to see if the Space Station itself will survive the domestic U.S. budget problems which have plagued it since inception.

According to the agreements signed, NASA has overall station-keeping management control of the station and its three pressurized laboratories, although module uses can be decided by multinational consensus. Each laboratory provider, all partners except Canada, can veto the use of its laboratory by another partner if for military purposes. The U.S. Defense Department still, however, has overall access to the station for national security work.

Why did the Europeans decide, after the long and sometimes strained negotiations, that participation was worth it in the long run? Although the commercial argument still gets first billing, it is recognized in Europe as in the United States that commercial returns are a long-term venture. Two factors of a more immediate nature combined to keep up the requisite momentum of support. The first is that Columbus is a kind of "logical extension" of past European projects, and one which will allow Europeans not only to develop their own space infrastructure, but also to position themselves for future economic returns. Second, there is the ever-present-in-cooperative-ventures political factor. The hope is that the Space Station will forge not only political links, but also links which will enable and encourage expansion of high technology industries across a broad front.

From the U.S. perspective, the Space Station capabilities will be much enhanced by European, Japanese, and Canadian participation. More practically, as in the past, NASA is hoping to use the "international commitment" aspect of the Space Station to protect it from devastating domestic budget cuts. It appears, however, that may not be the case. In September 1989 NASA initiated a major review of the station program to identify areas that could be reduced, depending on the health of the project's FY 1990 budget. The question then becomes: will the "commitment" of the international partners transcend the practical consequences of further station cut backs and almost certain launch delays for their modules? All parties may well be put to the test of living up to their "partnership."

Endnotes

1. ESA/C–M/LXVII/Res. 1 (Final), p. 3.
2. See: European Space Agency. Council, Resolutions adopted during the meeting at the ministerial level held in Rome on 30–31 January 1985, ESA/C(85)13, Paris, 4 February 1985.
3. It was not only the process of the Spacelab negotiations or the resultant agreement which was a source of concern, but for some Europeans, the implementation of the agreement as well. This group has in fact felt that the United States failed to live up to its obligation not to duplicate Spacelab technology in the United States.
4. "Europeans Hesitate on U.S. Space Station Plan," *Aviation Week & Space Technology*, 18 June 1984, p. 24.
5. See also: "After Spacelab, Europe Wants a Better Deal," *Science*, 9 December 1983, pp. 1099–1100; "Europe's Station Participation Viewed as Step Toward Own Facility," *AWST*, 3 June 1985, pp. 149–151.
6. Dr. Hermann Strub, "Columbus—The European Contribution to the International Space Station," Proceedings of the International Symposium on Europe in Space—the Manned Space System—held in Strasbourg, France, 25–29 April 1988 (ESA SP-277, October 1988), pp. 50–51.
7. Wulf von Kries, "Flunking on Space Station Cooperation?" *Space Policy*, February 1987, p. 11.
8. See, for example, Michael Feazel, "Europe Pushes Space Station Role," *AWST*, 18 June 1984, p. 16.
9. "Europe's Station Participation Viewed as Step Toward Own Facility," *AWST*, 3 June 1985, p. 149.
10. Multiyear authorization cycles for space-related programs have, in fact, received some support from Capitol Hill. See: Congressman Robert A. Roe, "Funding Long-Term Space Objectives," *Space Times*, July–August 1988, pp. 9–11.
11. U.S. Congress, Office of Technology Assessment, *Space Stations and the Law; Selected Legal Issues—Background Paper*, OTA-BP-ISC-41 (Washington, D.C.: U.S. Government Printing Office, August 1986).
12. Ibid., p. 53.
13. Ibid., p. 30.
14. Michael Feazel, "Germany Cites Commercial Fallout as Justification for U.S. Station Involvement," *Commercial Space*, Spring 1985, p. 47.
15. Ibid., p. 54.
16. See: Nathan C. Goldman, *American Space Law: International and Domestic* (Ames, Iowa: Iowa State University Press, 1988), p. 143.
17. DOD had never been in favor of the doctrine and had continually encouraged NASA to drop it.
18. John Nobel Wilford, "Pentagon Wants Space Station for Missile Study," *The New York Times*, 29 December 1986, p. 1. This turn of events was particularly surprising and disturbing because in 1983 Secretary of Defence Casper Weinberger had signed a memorandum stating that DOD had identified no national security requirements for the Space Station. See: Wayne C. Thompson, "West Germany's Space Program and the European Effort," *Space: National Programs and International Cooperation* (Boulder, Col.: Westview Press, 1989), p. 55.
19. The letter is referenced, for example, in an editorial in *AWST*, 20 April 1987, p. 11.
20. NASA Advisory Council, Task Force on International Relations in Space, *International Space Policy for the 1990s and Beyond*, October 12, 1987, p. 30.

Chapter 14

Space Applications

Earth Systems Sciences

Beyond exploration, beyond technology, beyond the general quest for knowledge, and even beyond profits, what relevance does space have for individuals living on Earth today? This often asked question was addressed in a recent issue of *Earthquest,* in relation to a discussion of revolutionary scientific advances. It is pointed out in the article that:

> Historically, revolutionary scientific advances follow within decades or centuries the invention of a new instrument—or family of instruments—for bringing under observation that part of the universe scientists wish to study. The telescope led to revolutionary advances in research on the solar system and other galaxies; the microscope and its descendants paved the way for molecular biology.
>
> In a narrow sense, the earth sciences do not have a revolutionary new tool. In a broader sense, however, remote sensing from spacecraft, coupled with modern communications and computer technology, are beginning to bring within reach a powerful new and different way of viewing the biosphere and geosphere. The attributes of this potential observing capability are uniquely attuned to the growing perception of planet earth's environment as a closely linked array of dynamic physical, chemical, biological, and social systems, a symbiosis that profoundly affects earth's habitability.[1]

What can be more relevant than a field through which humans are better able to understand the physical, chemical, biological, and social interactions which concern the Earth's habitability?

The imperative for developing capabilities to better understand our own planet is clear. Since the Industrial Revolution 200 years ago, the combustion of fossil fuels has increased the amount of carbon dioxide in the atmosphere by 25 percent. This, combined with the depletion of the equatorial rain forests which have been replenishing the oxygen in our atmosphere, has been found to be a problem global in scope and exemplary of other environmental problems Earth inhabitants will face in the not-too-distant future. A recent article in the publication *Interavia Space Markets* outlines the situation.

> Many experts predict the climate will revert to conditions last found in the geological record hundreds of thousands of years ago. These were radically different from the environment in which major civilizations have bloomed over the past 5000 years. Potential effects include a shift in climate zones—playing havoc with the distribution of species and agricultural productivity of North America and the U.S.S.R.—and a rise in mean sea level to partial melting of the polar icecaps. The latter event could cause the flooding of 80% of the U.S. coastal wetlands, likewise heavily populated regions from England to Bangladesh, and complete obliteration of low-lying islands in the Pacific and Indian Oceans. Suggested timescales range from decades to centuries, but most experts agree that major changes in the global climate are inevitable by the mid-21st century.[2]

Clearly it would more than behoove everyone to be able to at least accurately determine the situation the Earth is facing, rather than make speculations based on less-than-complete data.

Remote sensing, or collecting data about the Earth and its atmosphere from space for survey and monitoring purposes, is not a new venture. The political, social, and economic benefits of using space technology to look back at and study the Earth have long been recognized. American weather satellites were the first systematic Earth remote sensing spacecrafts, beginning to broadcast crude television images of cloud patterns in 1960.[3] The U.S. Landsat land remote sensing program has indeed been among the largest of NASA's international programs to date, involving researchers in forty-six countries and four international organizations with Landsat ground stations in twelve countries. It has been in operation since 1972—with major foreign policy implications.

A brief examination of the Landsat experience, in terms of remote sensing as a public concern versus remote sensing as a potentially lucrative commercial venture, is basic to considering future cooperative remote sensing ventures.

Landsat

Remote sensing from space for civilian purposes began with a NASA program initiated in 1964.[4] The first of five Landsat satellites—then called the Earth Resources Technology Satellite (ERTS-1)—was launched in 1972; Landsat 5, the last in the series to date, was launched in March 1984, after Landsat 4 began to fail. The Landsat satellites follow a polar orbit which takes them on a path covering the same spot on Earth at the same time of day, every sixteen days.

Each of the Landsat spacecrafts has carried a multispectral scanner (MSS) which has a spatial resolution of 80 meters and senses in four spectral bands. The Landsat 5 spacecraft has evolved to carry an MSS and a thematic mapper (TM) sensor having a spatial resolution of 30 meters[5] and seven spectral bands. Data are transmitted in digital form from the spacecraft to the ground stations, collected on tape, corrected to remove distortions, and prepared for sale as either photographs or on computer compatible tapes suitable for additional processing by large computers.[6]

From the beginning of the Landsat program, data was sold to anyone in the world, at reasonable prices. The data proved valuable for a wide range of applications, including improved mapping capabilities, resource surveys, and harvest prediction. Yet although the data was useful, growth in data sales grew more slowly than had been expected. That was partially due to inadequate government programs to recruit potential new users. Also, because Landsat depended on government funds for support, and because U.S. government funds are so notoriously tenuous, potential users were reluctant to become dependent on data from a program with an uncertain future. This led to a spiral effect; when use was lower than expected, government support for funds decreased, discouraging users further, and so on.

During the Carter administration, Landsat became a candidate for privatization. Carter's desire to reduce the size of the federal government coupled very well with the notion that enterprises such as Landsat with an actual product for sale could be run more effectively by the private sector. The same premises were supported and pushed even harder by the Reagan administration—during which time the Office of Management and Budget basically promoted the idea of privatization of government concerns wherever remotely feasible to cut down government spending. This was not the first time talks on a privately owned Earth resources venture had come about. Private industry had always maintained the position, however, that in order to make the venture profitable, all data gathered by satellite would be proprietary and sold to the highest bidder, rather than available to anyone. This, of course, raised a variety of questions ranging

from commercial exploitation of third world countries to whether or not a private concern could claim proprietary rights over photographs taken of countries without their consent.[7]

But the basic debate surrounding the privatization of Landsat was more fundamental—are the services provided by Landsat in the public interest and hence the system more appropriately operated by the government, or are the Landsat services actually a business like any other, and therefore more appropriately operated by the private sector? M. Mitchell Waldrop has characterized the problem as one being caused by the system's own success.

> The reason Landsat was so controversial, ultimately, was that the system was so successful: the data were incredibly useful. They could be used for science. They could be used for public purposes such as crop forecasting, Sierra snowpack monitoring. And they could be used for commercial purposes such as oil and mineral prospecting.
>
> Yet that was precisely the problem. It was all the same data. Thus, there was an immediate and inevitable controversy over who was going to operate the system. On the one hand, we had scientists and various government agencies saying that land remote sensing should be a government responsibility like the weather satellites. The vast majority of uses were for the public good, they argued, and anyway, there was not a big enough market to support a private system. On the other hand, we had would-be entrepreneurs screaming that the government had to get out of Landsat operations because it was stifling the development of a private remote sensing business.[8]

Also, beyond the more traditional "public good" type uses of Landsat data, which have foreign policy implications because of the goodwill that can be generated from sales abroad which assist developing countries, data has also been used for purposes with more direct foreign policy implications. For example, satellite images showing evidence of cocaine production have been used to fight drug trafficking.

But the decision had been made to privatize. Transferring Landsat from the public to the private sector, even with the vigorous attitude of the Reagan administration, became one of the most painful and prolonged processes taken on by the federal government in an effort to lighten its workload and budget commitments.[9]

Landsat was transferred from NASA to the National Oceanic and Atmospheric (NOAA), to be run as an operational system along with the national weather satellites operated by NOAA. In 1984, the Land Remote Sensing Commercialization Act of 1984 (P.L. 98–365) was passed, requiring that all U.S. firms involved in remote sensing provide nondiscriminatory access to data so that it is available at the same time and price to all users. A private company, the Earth Observation Satellite Company (EOSAT),[10] received a government contract in 1985 to privatize the Landsat system. What that means is that the government still owns the satellites and pays for their operation (through NOAA), but gives the data to the EOSAT company to sell the images. Problems, however, were quick to surface.

Factors which have inhibited the success of this venture include:

- Increased prices after privatization (EOSAT quadrupled the price of standard images)
- The dictate of the 1984 Commercialization Act that exclusive rights to data cannot be given to any one customer (with a consequent decrease in the value of the data from a proprietary stance)
- Poor handling of the commercialization transfer by all involved parties
- Foreign competition

In 1989 Landsat was faced with a shutdown when NOAA said it could not continue operation of the satellites after March 31 due to lack of funds. The funding problems arose because despite the government agreement with EOSAT for a reducing subsidy, in each of the previous three years the president had not requested any funds for the Landsat system, and it had been up to Congress to step in and provide funds. In 1989, however, Congress only provided funding through March. In a last minute reprieve the National Space Council, headed by Vice-President Quayle, announced that emergency funding would be provided to keep the operating Landsat 4 and 5 satellites orbiting while the Council conducted a full review of the Landsat program.[11]

Landsat has been a victim not only of its own success, but also of the typical U.S. unwillingness to define the appropriate role of government versus the private sector, coupled with the assumption which has been made in all but a very few cases (e.g., COMSAT) that they are mutually exclusive. Landsat has been monumentally important, however, in pointing out the advantages of using spacecraft to observe the Earth. Other ventures—national, multinational, private, and public—have been quick to follow.

International Remote Sensing Activities

Within the realm of space services, remote sensing has been viewed perhaps the most optimistically as having the potential for lucrative profits in the near term, if activity can indeed be used as an indicator of expected returns. More likely, activity has been spurred by estimates of potential return. For example, in 1985 the Center for Space Policy, a Cambridge, Massachusetts, think tank, forecast that gross revenues for remote sensing could reach $2 billion annually by the year 2000.[12] The problem is that although there are extremely high benefits to society in general from the use of land remote sensing data, it has been difficult to transfer or provide those benefits in terms of private goods in the marketplace.[13] Although the $2 billion figure has been shown to be an overestimation, the utility of remote sensing—if not the estimated potential income—has been recognized.

At least twenty-four countries or multinational organizations have known satellite Earth observation interests and capabilities.[14] Not all of these are by any means in the commercial field—yet—and it is still unclear what the real market potential will be.[15]

Several competitive foreign land remote sensing systems are currently operational or planned for deployment. They include:

1. *West Germany.* Modular Opto-electronic Multispectral Scanner (MOMS). Developed by West Germany and first flown aboard Shuttle Flight 7. There were high hopes for use of MOMS with the SPAS platform as a commercial enterprise, but they were quashed due to incompatibility with U.S. regulations. (See Chapter 8: SPARX.)
2. *France.* System Probatoire d'Observation de la Terre (SPOT). Developed through the French Space Agency CNES, this venture has been planned since 1978 as the world's first commercial remote sensing satellite service. Marketing is done through the company SPOT IMAGE S.A., which is partially owned by the French government. Although the SPOT system has higher resolution (10 meter) than possible with Landsat, as well as the capacity to return quasi-stereo data to the user, it also has fewer spectral bands, an important consideration for comparative purposes with other systems. In 1989, France approved the development of a fourth satellite as part of the SPOT system, to assure continuation of commercial services into the next century.
3. *India.* IRS-1. Built by India as a low resolution remote sensing system, it will be launched by a Soviet launcher.
4. *Japan.* Earth Resources Satellite (JERS-1). Designed for launch on an H-1 vehicle in 1991, JERS-1 will have both oceanic and land applications. The primary mission of this system is gathering information on renewable and nonrenewable national resources. Japan has also built a marine observation satellite (MOS-1), launched in February 1987. The satellite carries microwave, multispectral, thermal, infrared, and visible radiometers, including detectors comparable to those on the French SPOT system.
5. *Brazil.* Work is being conducted on a moderate resolution land-sensing satellite for launch in the late 1980s.[16]

6. *China.* China is considering setting up its own Earth observation satellite systems: the first Chinese Meteosat is to be launched into a polar orbit as part of the Seventh Chinese Five-Year Plan.[17]

7. *Soviet Union.* Soviet efforts, including the Kosmos series of satellites in LEO, are directed toward Earth resources photography, with 5 meter resolution. The Soviets also make extensive use of their manned space stations for Earth observation. MIR is used for a routine program of multispectral photography, as had been the case with earlier Salyut spacecrafts. Specialized satellites in LEO, such as in the Meteor and Okean series, are also used for remote sensing. The Soviets began actively marketing these pictures worldwide through their organization Soyuzkarta in 1987.

Thus, Landsat, SPOT, IRS-1, and JERS-1 all have land remote sensing capabilities. Multiple countries are also operating polar and geostationary weather satellites, as well as oceanic satellites. The obvious lack of regard for national boundaries of these satellites, coupled with their clear utility in monitoring global change, has prompted some interesting proposals and projects which perhaps indicate a future wave of international cooperation in space.

Mission to Planet Earth

Whereas issues such as the greenhouse effect, ozone depletion, acid rain, deforestation, and drought were once confined to discussions among scientists, that is no longer the case. Not only have these issues gained the attention of politicians worldwide, but the public is concerned as well. In fact, perhaps public interest preceded and prompted interest by politicians. In the United States, NASA has taken the lead in the effort to monitor Earth from space with a proposed $15–30 billion project that would extend over fifteen years, called Mission to Planet Earth (MPE). Suggested in the 1987 report of a commission headed by former astronaut Sally Ride, Mission to Planet Earth has been strongly supported by the Bush administration. The purpose of this mission will be to monitor global environmental change and document the environmental interactions among air, land, and sea environmental forces.

Specifically, the program involves a series of spacecraft, with launches to begin in 1996. Minimally, the spacecraft program will require $15 billion spread over fifteen to twenty years. The most ambitious programs plans, if approved, have been conservatively estimated as requiring approximately $30 billion. Initial plans have been approved and funded. What scale the program will finally take on is basically up to the commitment and endorsement finally given by the Bush administration. If the smaller scale plans are implemented, NASA will launch four to six unmanned space platforms, weighing approximately 15 tons each, into polar orbit. The launches, using Titan 4 boosters, would span the period of 1996 to 2011. Two of the NASA polar orbiting platforms would be part of the Space Station. In addition, small satellites would be launched into equatorial orbits to obtain additional, specialized data. For a full-scale Mission to Planet Earth to be implemented, most experts agree that a group of five geosynchronous satellites will be required.[18] If these geosynchronous spacecrafts were included in the U.S. plans, costs would be about $2 billion per year.[19] MPE faces its strongest competition for funding from major astronomical and planetary programs currently planned by NASA. Careful planning to avoid peak funding periods coming at the same time could alleviate some of those problems.[20]

Mission to Planet Earth is an attractive mission politically, economically, and scientifically. Within the United States, MPE is particularly attractive politically because it includes a focal role for the Space Station, which in turn expands the Space Station list of practical uses. It is also attractive because of the international character it has assumed. With recognition of the extent and seriousness of the issues that Mission to Planet Earth will deal with, has come recognition of the inherently global nature of those problems. Coordination of efforts has become not only a national, but a multinational effort. The reason is simple; the science is not just important, it is imperative.

Besides the United States, Europe and Japan are already involved with Mission to Planet Earth, and the Soviets have expressed interest in joining the effort also. Both Japan and ESA will launch at least one polar orbit platform each during the intended period of study and have spacecraft plans already in the advanced definition phase. Plans for the Soviet Union to join with the United States to monitor changes in the Earth environment have focused on the use of separate but complementary hardware. Beyond the scientific merits, the viability of such a proposal is high because of each country's ability to guard its own technology and because this is a noncommercial and hence noncompetitive venture.[21]

Although Mission to Planet Earth is still in the planning phases, steps have already been taken to coordinate Earth-viewing satellites in the near term. Space officials from twenty-three countries are coordinating the operation of over twenty satellites to be launched over the next six years. Originally intended to operate separately, they are now being coordinated as part of the 1992 International Space Year (ISY), which is being held to commemorate the five hundredth anniversary of Columbus' discovery of America. ISY officials have adopted Mission to Planet Earth as a major theme, as well as assisting with the broader International Geosphere-Biosphere Program.

International Geosphere-Biosphere Program

Acknowledging that the study of the Earth is an inherently multinational endeavor, the International Council of Scientific Unions (ICSU) endorsed an International Geosphere-Biosphere Program (IGBP) in 1986, intended to examine the processes that affect the Earth as a system and the subsequent role of human activities. A special committee for the IGBP met in Stockholm in October 1988 to set the fundamental goals and guidelines of the program. Within the United States, the National Academy of Sciences has established a Special Committee for the Global Change Program (GCP). This committee works closely with the interagency Committee on Earth Sciences (CES) within the federal government. Many of the U.S. contributions to the IGBP will be drawn from a broad range of ongoing U.S. observational and research programs.

Within the United States, coordination of efforts led by Mission to Planet Earth are planned within the U.S. Global Change Program (GCP). This program is to encompass the global geosciences study at the National Science Foundation, Earth System Science/Mission to Planet Earth at NASA, and the Climate and Global Change Program at NOAA. The program has some very strong pros and cons in its political and scientific favor. Coordination of efforts maximizes the scientific return yielded from a project, as well as providing agency planners with priorities and guidance for the future. Also, since several agencies are involved in the program, multiple parties have a stake in making it work, strengthening the program's position in budget debates. However, it will still be subject to the political and budget imperatives that every other program is subjected to. Perhaps most detrimental to the arguments supporting the program before budget critics is that it offers no tangible product. It offers data, and the data will doubtlessly offer insight and trigger ideas, but there will be nothing else upon which to judge its success or failure. This makes it critical that NASA institute a public awareness campaign to make the program benefits—the information interpreted from the data and how it is subsequently used—known to the public. The factor which may in fact maintain the political balance of support in favor of the Mission to Planet Earth/U.S. Global Change Program is its international aspects.

Although it is widely acknowledged that the International Council of Scientific Unions provides a well-established framework for developing the substantive scientific program for Global Change Program, the Council is limited in its operational capabilities. It has therefore been suggested that:

> Given the unique characteristics of the GCP, it seems clear that another organization needs to be enlisted (or created) to bring together the national and international agencies that will manage and fund the programs on which the GCP will be built.
>
> An umbrella operational entity will be necessary for several reasons. First of all, an international data system will need to be designed and managed not for the length of a single scientific project but for decades. A forum will also be required for development and management of bilateral and multilateral research projects in a wide variety of disciplines. And finally, of course, there will be the issue of funding—not only funding of research projects by government agencies in the developed world, but funding of training centers, ground-based data collection networks, and research projects in the developing world.[22]

Such an entity has been proposed.

ENVIROSAT

The concept of an international Environmental Resources Satellite (ENVIROSAT) consortium to unite the technical strengths and market needs of countries involved in remote sensing has been discussed in various forms since the early 1980s. At first, discussions focused on a COMSAT initiative called EarthSTAR, proposed to combine the U.S. land and weather satellite systems under a commercial owner/operator. The owner/operator would sell the land remote sensing data on the open market, but would have an exclusive contract with the U.S. government to provide the National Weather Service with all the meteorological data acquired. Beginning in January 1985, John L. McLucas presented a series of editorials in the magazine *Aerospace America*,[23] urging an international approach to remote sensing. From there, discussions were held at several international space meetings and forums, gathering support for the concept. John H. McElroy, in a statement before Congress in 1987,[24] coined the name ENVIROSAT, which was then more or less adopted for the effort by consensus. INTELSAT and INMARSAT are often proposed as models for any ENVIROSAT organization. Drawing from the INTELSAT/COMSAT relationship as an example, it has been suggested that ENVIROSAT-International would be responsible for providing a core set of public and private data to participating nations and licensees. ENVIROSAT-USA would be responsible for consolidating the facilities that would replace the separate, dedicated facilities of NASA, NOAA, and EOSAT.[25] Although the details remain to be worked out, the premise is clear, and increasingly attractive to many policy makers.

Remote Sensing in the Future

Experience has brought forth many important and illustrative points for consideration of future environmental remote sensing activities. The satellite activity involved with gathering information about the Earth and its environment is inherently global in nature, as is the data being gathered and the problems to be dealt with through the data. The problems are recognized as being not only important, but urgent. With limited budgets a way of life for most nations, the field naturally lends itself to cooperative efforts.

But should remote sensing activities be public or private? To be the most desirable and efficient, private sector management must be done economically and still meet political needs and provide data on an open basis. Assuming all this, there is much to learn from the Landsat experience. For example, for any remote sensing venture to be successful it minimally must be stable, provide data at reasonable prices, and have an assured market. Landsat experience showed that without stability customers will be hesitant about becoming reliant on the services offered. Assuming that open access to data remains a policy for future ventures, "public good" type data will be the prevalent data purchased. Some of those purchases will be on an occasional or ad hoc basis and therefore will be purchased only if the price is maintained at an affordable level. Any organization dependent on these sales alone, however, would find it difficult to prosper economically. Therefore, maintenance of a larger, dependable market must also be assured. Being the data supplier for national weather services would give that organization the assured market necessary for it to be profitable and hence able to expand, much as INTELSAT and INMARSAT have.

Like many other types of cooperative efforts in this third period of study, remote sensing is an example of an open-ended program. These types of projects have all been fundamentally lacking in infrastructures necessary for long-term management. An international organization is indeed needed for this purpose, and the sooner it is in place and operating the sooner countries will be able to maximize not only their available funds, but their returns on investment—in terms of data yielded—as well.

Conclusion

Earth system sciences is an area especially ripe for expanded cooperative efforts because activity is already underway, being funded, and recognized as imperative, subsequently making cooperation a question of degree and scope. By the time of the ISY in 1992 and the concurrent commencement of Mission to Planet Earth, using spacecraft for the study of our own planet shows every indication of being the dominant area of cooperation in space. Planners involved with MPE have drawn from the most successful parts of other cooperative programs as models, trying to avoid potential pitfalls wherever possible—a trend which hopefully will continue and grow.

Endnotes

1. Thomas F. Malone, "Global Change—Paradigm, Progress, Process," *Earthquest*, Spring 1989, p. 2.
2. Stephane Chenard, "Apocalypse When?" *Interavia Space Markets*, May/June 1989, p. 71.
3. *Space-Based Remote-sensing of the Earth: A Report to the Congress* (Washington, D.C.: NOAA of the U.S. Department of Commerce and the National Aeronautics & Space Administration, September 1987).
4. See: Pamela E. Mack, "Space Science for Applications: The History of Landsat," in *Space Science Comes of Age* (Washington, D.C.: Smithsonian Institution Press, 1981).
5. Except for the 10.4 to 12.5 micron wavelength band which has a spatial resolution of 120 meters.
6. *International Cooperation and Competition in Civilian Space Activities* (Washington, D.C.: U.S. Congress, Office of Technology Assessment, OTA-ISC-239, July 1985), pp. 278–280.
7. The latter issue had especially sensitive political overtones, as many people felt it could result in pressure for international laws that wold restrict the taking of photographic images from space and subsequently also affect the use of reconnaissance satellites—any legal restrictions on which have always been scrupulously avoided. For a discussion of some of the legal issues involved with remote sensing, see: Glenn H. Reynolds and Robert Merges, *Outer Space: Problems of Law and Policy* (Boulder, Col.: Westview Press, 1989), pp. 178–194.
8. M. Mitchell Waldrop, "The U.S. Global Change Program—A Political Perspective," *Earthquest*, Spring 1989, p. 6.
9. Paul Mann, "Remote Sensing: A Tortuous Trip to Marketplace," *Commercial Space*, Spring, 1985, pp. 32–37.
10. A joint venture of RCA Corporation and Hughes Aircraft.
11. "Death Postponed for Landsat Satellites," *Science News*, 18 March 1989, p. 172.
12. Mann, "Remote Sensing: A Tortuous Trip to Marketplace," p. 32.
13. John Logsdon and Tracie Monk, "Remote Sensing from Space: A Continuing Legal and Policy Issue," *Annals of Air & Space Law*, Vol. VIII, 1983, p. 412.
14. They are: Argentina, Australia, Belgium, Brazil, Canada, EEC, ESA, FRG, France, India, Italy, Japan, Netherlands, Norway, Pakistan, PRC, Republic of South Africa, Saudi Arabia, Sweden, Switzerland, Thailand, U.K., U.S., U.S.S.R., per John L. McLucas and M. Maughan, "The Case for Envirosat," *Space Policy*, August 1988, pp. 229–239.
15. One interesting use of satellite imagery gathered from space is use by the news media. See: Richard Dalbello and Ray A. Williamson, "Gathering News from Space," *Space Policy*, November 1987, pp. 298–306.

16. *International Cooperation and Competition in Civilian Space Activities*, pp. 282–285.
17. See: He Changchui, "The Development of Remote Sensing in China," *Space Policy*, February 1989, pp. 65–74.
18. The United States would supply two, with the others provided by Europe, Japan, and possibly the Soviets.
19. Craig Covault, "Major Space Effort Mobilized to Blunt Environmental Threat, *Aviation Week & Space Technology*, March 13, 1989, p. 44.
20. See: Chenard, "Apocalypse When?" p. 72.
21. Paul Adam Blanchard, "Combined Remote Sensing Observations of the Earth from Space," Foreign Policy Institute, SAIS, Johns Hopkins University, November 1988.
22. Robert Palmer, "Making the Global Change Program a Global Research Program," *Earthquest*, Spring 1989, p. 9.
23. John L. McLucas, "Whither Landsat," *Aerospace America*, January 1985; "Open Skies: A Fresh Challenge," *AA*, April 1985; "Space, Sanctuary or Menace?" *AA*, May 1985; "Create a Global Landsat," *AA*, July 1987.
24. John H. McElroy, Statement on *The Future of Landsat*, before the Subcommittees on Natural Resources, Agricultural Research and Environment and Subcommittee on International Scientific Cooperation, Committee on Science and Technology, U.S. House of Representatives, 2 April 1987.
25. For detailed examinations of ENVIROSAT proposals See: John H. McElroy, Jennifer Clapp, and Joan C. Hock, "Earth Observations: Technology, Economics, and International Cooperation," *Economics and Technology in U.S. Space Policy*, Molly K. Macauley, ed. (Washington, D.C.: Resources for the Future, 1987), pp. 25–44; John L. McLucas and Paul M. Maughan, "The Case for ENVIROSAT," *Space Policy*, August 1988, pp. 229–239.

Chapter 15

A New Model of International Cooperation: The Inter-Agency Consultative Group[1]

Introduction

Kenneth S. Pedersen, former NASA director of international affairs and currently associate administrator for external relations, has written on the problems inherent in the new environment of cooperation in space, and the subsequent issues associated with "balancing national or regional pride with the fiscal and technical imperatives pushing towards broader-based multinational cooperation."[2] Pedersen has suggested a "confederal" approach, whereby countries develop discrete, national hardware, which is then coordinated to optimize utility, as one way to deal with that problem. This approach was in fact employed in the multinational study of Comet Halley in 1986, by a group known as the Inter-Agency Consultative Group (IACG), formed specifically for the purpose of coordinating national efforts.

The IACG is an example of recent international attempts to acknowledge the need for cooperative efforts and to find a solution to the dilemma so succinctly described by Pedersen of balancing national and international interests. The United States has been noticeably reluctant of late to become involved with new transnational space institutions, an attitude advocated by the 1986 National Commission on Space report recommendation "against U.S. support for any global organization that purports to regulate broadly the utilization of outer space."[3] This reluctance is not necessarily shared by other countries. Proposals have been made by the French to establish an international satellite monitoring agency, the Europeans have already established a history of cooperation encompassing the European Space Agency (ESA), Arianespace and SPOT Image, and the Soviets have proposed the establishment of a World Space Agency. Although the latter is a clear effort to inhibit the U.S. Strategic Defense Initiative, and the Soviets do not want any type of organization that can impinge on their activities any more than the United States does, it is nevertheless indicative of the trend toward increasingly multinational cooperation.

Pedersen, in an article from a 1986 symposium presentation, raises the notion of an International Space Science Agency (ISSA) to further cooperation in that field. He states that space science is perhaps the most feasible area for initiating cooperation on an institutional basis for a variety of reasons, including associated project costs, a history of collaboration between scientists, and the ability to minimize technology transfer problems. He says that "Logic suggests that a new international institution such as an ISSA would have to be the culmination of a series of incremental steps. One such step may be currently underway."[4] The reference there is to the IACG. Pedersen goes on to say that:

> Assuming that the IACG or a similar organization continues to perform successfully, steps could be taken to formalize its existence through a multilateral agreement. Consideration could later be given to expanding the group's membership and endowing it with limited decisionmaking and financial authority as a possible prelude to assuming a more independent status.[5]

And finally, he raises the issue of "expanding the mandate of the IACG to include coordination of other areas of mutual interests—for example, in the exploration of Mars."[6]

The IACG has completed its first mission, the coordinated study of Comet Halley, and is involved with its second project, Solar Terrestrial Science. Therefore the time is propitious to review the activities of the group in the context of the issues raised by Pedersen, and evaluate the IACG as an incremental step for expanded multinational cooperative space efforts in the future.

Preparing for Comet Halley

The United States was the only major space power not to send a dedicated spacecraft to Halley's Comet in 1986 as part of an international armada. The decision by the United States not to build and send a spacecraft to Halley's Comet was not based on a merit evaluation of the project, but was more the result of internal politics, with higher scientific priority placed on other missions being a key factor, especially in the early stages of the decision-making process.[7] Some officials involved in the decision, however, maintain that the decision was ultimately made on budgetary grounds. A Halley spacecraft would have had to compete with projects like the Gamma Ray Observatory and Galileo for funds, and those were given higher priority. NASA was still to be a participant in the Halley's armada, however, its part being to fly a telescope on the Shuttle for comet observation (which of course it was unable to do), and to observe the comet with existing spacecraft such as the International Ultraviolet Explorer (IUE), the Solar Maximum Mission (SMM), Dynamics Explorer (DE) I, Pioneer 7, Pioneer-Venus Orbiter (PVO), and the International Comet Explorer (ICE).[8]

The U.S.S.R., ESA, and Japan all had programs which included one or two dedicated spacecraft to observe the Comet Halley. The Soviet agency Intercosmos sent two spacecraft, Vega 1 and 2. The Japanese Institute of Space and Astronautical Science (ISAS) also sent two spacecraft, designated Sakigake and Suisei. The European Space Agency (ESA) sent one spacecraft, called Giotto. It was recognized early on that scientific returns from these individual missions would best be maximized by some form of coordination of efforts.

The Inception of the IACG

A scientific event of the magnitude and infrequency of Halley's Comet necessarily draws preparatory attention from the scientific community for many years before it actually occurs. Early in the preparation for the return of Comet Halley some involved parties, including French scientist Jacques Blamont, suggested that coordination of national efforts could be accomplished through the Committee on Space Research, COSPAR. However, other scientists and space officials felt that it would be better if space agencies could work directly with each other, rather than working through another organization, particularly a nongovernmental organization which would ultimately have no authority or ability to carry through its plans. To realize that goal, the IACG was created.

Credit for the initial formation of the IACG should be given to Drs. E.A. Trendelenburg, the ESA director of scientific programs from 1975 to 1983, and Roald Sagdeev, director of the Soviet Space Research Institute (IKI) from 1973 to 1988. Trendelenburg and Sagdeev both supported the idea that space agencies having Halley's missions should work directly with each other to maximize scientific return. Trendelenburg flew to Moscow in early 1981 for a one day meeting at the airport, planned specifically to discuss ideas for an informal coordinating mechanism. The fruition of

that discussion was the first meeting of space agency representatives concerned with Halley's Comet projects, held in September 1981.

Following the 1981 meeting of the International Astronautical Federation (IAF) in Rome, an annual meeting heavily attended by the type of space officials who could provide the impetus for organizing a new coordinating mechanism, Dr. Trendelenburg invited selected space agency representatives to Padua, a short distance away.[9] The first IACG meeting had delegations from NASA, ESA, ISAS, and Intercosmos, as well as representatives from the International Halley Watch (IHW). IHW was a group of professional astronomers, over one thousand in fifty-four countries, brought together primarily through the efforts of Ray Newburn at the Jet Propulsion Laboratory (JPL), with German scientists Jurgen Rahe coordinating European efforts, for ground observation of Comet Halley.

An Italian astronomer and long-time advocate of international scientific cooperation at the University of Padua, Giusseppe Colombo, hosted the first meeting. In his welcoming address, he spoke of space in a visionary sense. He challenged those present to go beyond their national interests for more than enhancing the data return on a comet, but for the good of mankind. He said:

> ... If you can succeed in reaching agreement on optimum integration of the various space probes and the ground and space observations beyond national interests and within the bounds of fruitful cooperation, you can show the world how men of good will have an intrinsic ability to work together to establish the truth, to increase knowledge among men, and to foster a peaceful and better society.

His message touched at least a few in the audience, including Dr. Rudeger Reinhard, the ESA Giotto project scientist, who later saw to it that the speech was published in the *ESA Bulletin*.[10]

Reinhard had assisted Trendelenburg before the meeting with development of an agenda. As the Giotto project scientist he had initiated an international working group on studying the Halley environment (focusing on the dust particles in the atmosphere of the comet nucleus) in 1979, and subsequently offered it to Trendelenburg as an organizing mechanism for IACG work, along with suggesting two other working groups. Trendelenburg was called away from the IACG meeting after only a short time and turned the chair over to Reinhard. As everyone else had come prepared primarily to listen, rather than to offer proposals, the entire Reinhard agenda for creating working groups was easily adopted.

At this first meeting, general working principles were established, information was exchanged concerning the various national Halley's ventures, and three working groups formed: the Halley Environment Working Group, the Plasma Science Working Group, and the Spacecraft Navigation and Mission Optimisation Working Group. The role of these working groups was to make actual recommendations concerning coordination of flight projects or task allocation. The working arrangements of the groups were kept as simple as possible:

> To minimise travel time and expense and in view of the difficulty of arranging meetings with working-group members from the four agencies present, it was agreed from the outset:
> —to try to make as much use as possible of existing major conferences when arranging working-group meetings
> —to communicate results achieved by sub-groups in regional meetings to all other working group members who could not participate and to a representative of each agency
> —to encourage flexibility of membership depending on the interest and changes in the subject matter of the discussions and depending on the meeting place.[11]

Working groups such as these have become the core of the IACG.

There was some discussion at this first meeting as to whether it was appropriate for NASA to be an IACG member, since it had no approved, dedicated spacecraft program. Many people felt, however, that it would be unthinkable to do a space science venture of this magnitude without the experience and involvement of NASA personnel. The matter was tabled on the basis that NASA still had a Halley's project, the Halley's Intercept Mission (HIM), under consideration at the time. A few weeks later, when HIM was dropped, NASA was already a de facto, IACG participant.

Following the initial session in 1981, IACG meetings were held annually, rotating among the four agencies involved. Meeting structure was deliberately kept informal. Delegations consisted of between five to ten senior scientists, engineers, and managers, selected according to programmatic interest by the delegation head, who was usually a senior level agency official. Although there was no secretariat, Rudeger Reinhard has acted as the IACG executive secretary since the inception of the group because of his personal interest in its success. This task has involved preparing the meeting agenda in consultation with liaison offices at the involved agencies and carrying out day-to-day activities between the scheduled meetings. Reinhard is often referred to as the "corporate memory" of the IACG.

The 1985 IACG meeting in Washington, D.C.,[12] was to be the last meeting before the March 1986 Halley encounters. A pamphlet was issued at that time describing the major components and objectives of the IACG coordinated Halley armada. The pamphlet was particularly interesting in that it had the logos of all four participating space agencies and the IHW all on the cover, evidencing the cooperative and clearly unique nature of the group.[13]

The meeting itself was a good public relations opportunity for NASA, as the ICE spacecraft was just passing through the Comet Giaobini-Zinner, and the IACG participants were able to make observations from Goddard Space Flight Center. This was the first actual experience the group had had with a comet, and NASA was justifiably proud that it occurred through a NASA program.

The IACG and Project Halley

Coordinating observations of Comet Halley was the inaugural project of the IACG, indeed the initial raison d'être of the group. Thus an examination of the dynamics of the group and the people involved in the implementation of Halley observations is appropriate. Retrospectively, many of the scientists and officials who were involved with the Halley observations feel that a unique combination of people allowed a new type of entity like the IACG to be formed and be successful. The uniqueness of the group stems from the fact that the people who were involved with IACG for the duration of the Halley's mission were by no means strangers; rather they were international colleagues who had known each other and worked together on various projects for years, and they were all committed internationalists.

All of the IACG participants felt comfortable with each other. They were colleagues coming together on projects of mutual interest. Reinhard observed that at the working group meetings, "scientific arguments often prevailed across boundaries of delegations."[14] The confidence they shared in each other was essential as this was an innovative project which outsiders were watching for success or failure.

According to Reinhard, the participants in the IACG can be divided into two categories, of equal importance: senior level officials and members of the working groups. Reinhard says of all of them, "It is necessary for these people to believe in international cooperation for the whole concept of IACG to work."[15]

Clearly, without the approval and support of senior level space agency people, often acting as delegation heads, the working groups would be hindered in their efforts. This has never been the case within the IACG. Initially, Trendelenburg acted as the delegation head and senior official for ESA, and Sagdeev for Intercosmos. NASA Associate Administrators for Space Science and Applications Andrew Stofan and Jesse Moore were the NASA representatives for the first two years, respectively. The ISAS delegation was headed by Dr. Minoru Oda, with Dr. Kunio Hirao acting in his place intermittently. Of those, only Sagdeev acted as delegation head for his agency during the whole of the Halley project. Roger Bonnet took over from Trendelenburg in 1983 as both ESA scientific program director and subsequently IACG delegation head. Geoffrey Briggs from the NASA Office of Solar System Exploration was consistently present at all IACG meetings for NASA and acted as delegation head from 1983 to 1985. NASA Associate Administrator for Space Science and Applications Burton Edelson chaired the meeting in 1985 and acted as delegation head in 1986. At the time of the Halley encounters, the delegation heads were then Sagdeev, Edelson, Bonnet, and Oda.

Roald Sagdeev served as director of the Space Research Institute (IKI) of the Soviet Academy of Sciences between 1973 and 1988. His part in the creation of the IACG has already been stated. A physicist himself, Sagdeev is internationally recognized as a advocate of space science cooperation. Sagdeev continually charms his Western colleagues with his charismatic ability to address audiences and to not only understand Western humor, but use it as well.[16]

Edelson came to NASA in 1982 from COMSAT where, over the fourteen years that he was there, he recognized that international cooperation through INTELSAT allowed that organization to exceed the capabilities of any one country. Subsequently he became a strong advocate of international cooperation for appropriate science projects, and he brought that attitude with him to NASA. Edelson was the first NASA associate administrator to take a serious interest in the IACG, fortunately at a time when high-level support was critical for the continuance of the organization.

Roger Bonnet is another individual committed to international cooperation for reasons beyond the scientific and political benefits it can yield. Bonnet had worked with the other IACG people, particularly Edelson, on a variety of prior projects, and they were comfortable with each other's philosophy as well as working style. To illustrate just how innovative the IACG was at its inception, committed internationalist Bonnet was somewhat skeptical at the first meeting, which he attended as part of the ESA delegation, about whether it would work. His skepticism was soon replaced by commitment, however, as the group began functioning effectively.

Finally, Minoru Oda headed the delegation from ISAS. Dr. Oda has a very outgoing nature, a characteristic somewhat unusual in the Oriental cultures. Further, his background includes time spent at MIT, so he is comfortable with Western style and culture.

He too brought a strong belief in international cooperation to the IACG. Professor Kunio Hirao, who headed the ISAS delegation between 1982–85, shared Professor Oda's views on cooperation.

It has already been stated that Rudeger Reinhard has served as IACG executive secretary since the group's inception. During the early years there were times of considerable doubt that the organization could withstand the rigors of its mission, primarily from a political perspective. The difficulties of merely coordinating meetings between the five groups seemed at times more than participants were willing to endure, and some people within individual organizations sometimes considered the easier route of doing things on their own. Reinhard took on the role of troubleshooter and became the driving force behind the IACG more than once.

The working groups are the heart of the IACG, as it is here that the actual scientific coordination and cooperation occurs. Reinhard himself chaired Working Group 1 on the Halley's environment, Soviet scientist (and later successor to Sagdeev's position at IKI) A. A. Galeev chaired Group 2 on plasma physics, and JPL scientist Jim Dunne chaired Group 3, which was basically the catchall group to cover everything else.[17] They—with others such as David Dale from the European Space Research and Technology Center (ESTEC), and from the United States Geoffrey Briggs at NASA, Fred Scarf from TRW, and Ray Newburn, Frank Jordan, and Don Yeomans from JPL—were individuals taking active roles in the working groups and on project coordination.

For the actual encounters, most of the IACG delegations were in Moscow first, for the Vega flybys, and then moved to the European Space Operations Center (ESOC) in Darmstadt for the Giotto flyby.[18] The Moscow encounters were the real beginning of Soviet *glasnost* in their space program. Real-time images were projected on large screens, putting the Soviets out on a limb in front of international audiences, something they were not accustomed to doing. Galeev narrated the activity on the screens. At the first sight of the nucleus on the screen an atmosphere of excitement overtook the room. The group was clearly pleased and proud of their individual and combined efforts. The pride continued at ESOC for the Giotto observation, which had been the clear result of the combined efforts of IKI, NASA, and ESA, with the Pathfinder mission.

Pathfinder

The Pathfinder[19] concept was proposed at the first IACG meeting in Padua in 1981 by Trendelenburg and approved at the November 1984 IACG after the technical specifications and plans for the individual spacecraft became better known. Because of the degree of

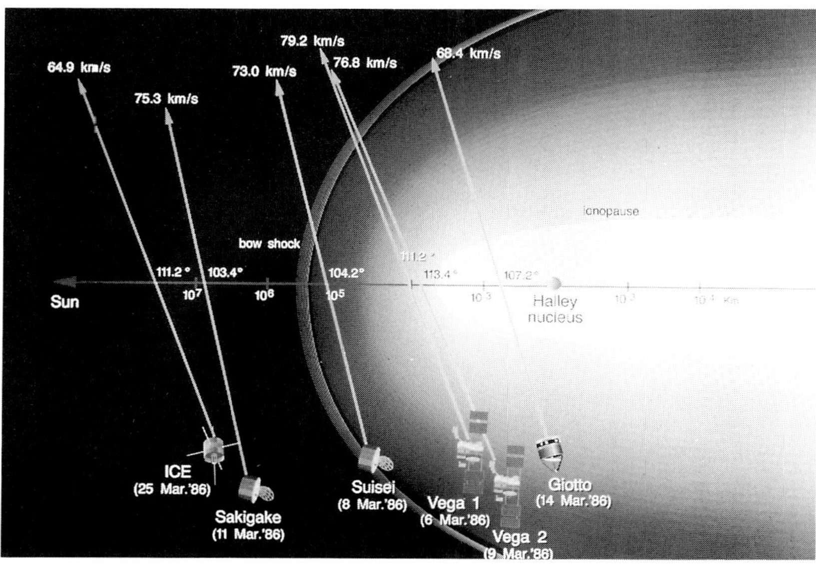

Figure 15.1 Spacecraft encounters with Comet Halley. *Photo courtesy of European Space Agency.*

technical coordination that would be required for Pathfinder to be successful, some people were originally skeptical, including David Dale, who chaired the steering committee for its technical implementation. But in the end he and virtually all others became committed to its success. Pathfinder basically involved using the two Vega spacecraft and the U.S. Deep Space Network to provide information required to target the ESA Giotto spacecraft as accurately as possible.[20] The two Soviet Vega crafts were the first spacecraft to arrive at the comet, on March 6 and 9, 1986, respectively.[21] Their cameras would detect the comet's nucleus and determine its position. The Deep Space Network, run through JPL, would carefully track the Vegas to enable calculation of a more accurate location of the Vega spacecraft. The Vega tracking data allowed course adjustments to be made on Giotto to target the craft with an accuracy of 40 kilometers. Giotto was targeted to encounter Halley at a distance of 540 kilometers, and the actual flyby occurred at a distance of 596 kilometers. Previous course estimates were accurate only up to 400 kilometers, which meant targeting with such accuracy would have been impossible without the coordinated efforts of the IKI-ESA-NASA project.

The success of Pathfinder depended upon efficient communication linkage between the agencies involved. Dr. R. Munch at the ESOC served as the technical coordinator for all of the agencies, making sure the organization, operations, and compatibility of math models, computer formats, and software, etc., were taken care of in advance. A permanent "hot line" was established between ESOC and IKI, and a computer-to-computer link was established between ESOC and JPL. But because of political considerations, all data

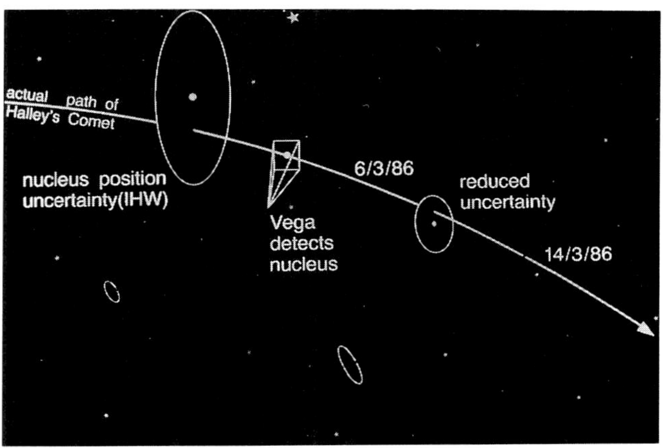

Figure 15.2 Pathfinder concept. *Photo courtesy of European Space Agency.*

between IKI and JPL had to be run through ESOC, since the United States and the Soviet Union had no agreement sanctioning cooperation in space at that time. The work of coordinating operations between IKI and JPL was, according to Dr. Munch,[22] the most difficult preparatory aspect of the project in terms of time required. The results, however, were spectacular.

The entire Halley's mission was a great scientific success. A symposium was held in October 1986 in Heidelberg, West Germany, to bring together over 520 participants to review the Halley's data.[23]

The IACG Beyond the Halley's Mission

The first "post-Halley" IACG meeting was to be held in November 1986 in Padua.[24] Burton Edelson had suggested at the 1985 meeting that permanent Terms

Figure 15.3   IACG delegation members meeting with the Pope after the successful Halley's Comet encounters. *Photo courtesy of European Space Agency.*

of Reference should be drafted for the IACG, and that the issue of whether or not to continue the IACG beyond the 1986 meeting (and if so what it should do) be put on the 1986 agenda. Some people within the IACG were hesitant about Edelson's suggestion, feeling that any degree of formalization had the potential to create problems for the group. But in 1985 Edelson took it upon the U.S. delegation to draft the Terms of Reference, which basically reflected in written form how the organization had worked over the past four years, and had this sent out in time for consideration at the 1986 meeting.

At the Padua meeting, the success of the missions and of the group generally were seen as compelling reasons for the organization to continue, leading to the unanimous decision to continue the IACG, now to be known as the Inter-Agency Consultative Group for Space Science. The Terms of Reference were finalized and accepted by the delegation heads. They were recognized as beneficial, in the sense of giving the group a certain degree of stability, and hopefully permanence, although delegation members still stressed the need for the group to remain as informal as possible, as part of their past formula for success.

The Terms of Reference clearly reflect the three

basic operating principles which, according to Reinhard, "everybody involved understands": there will be no or minimal technology transfer, no exchange of funds, and the group serves a strictly advisory function to the member agencies.[25] The purpose of the group is stated as:

> The objectives of the Inter-Agency Consultative Group for Space Science (IACG) are to maximise opportunities for multilateral scientific coordination among approved space science missions in areas of mutual interest. The IACG is a multi-agency international forum in which space science activities are discussed on an informal basis among representatives of member agencies.[26]

This statement of purpose carefully and succinctly states the boundaries within which the IACG operates. Without these boundaries, governmental support for the organization would have been difficult—or more probably impossible—to obtain, and hence the organization would not have survived.

Also included in the Terms of Reference are a set of policies to guide the operation of the IACG, and which in effect further define IACG boundaries within some very pragmatic guidelines.

—Where mutually agreed by the participating agencies, the

IACG serves as a vehicle for coordination efforts among approved space science missions on a multilateral basis.

—Exchanges of information on future plans and potential science missions are desirable, and take place during periodic IACG meetings. However, the IACG does not have a formal planning role for future missions.

—The IACG does not supplant bilateral cooperative space science activities and arrangements, nor does it serve as a substitute for existing mechanisms for managing specific multilateral space science projects.

—The IACG leadership (agency delegation heads) is comprised of senior space agency representatives in order to maintain the overall efficiency and productivity of the group.

—In addition, where appropriate, the IACG may consider the participation of organised ground-based scientific communities in order to enhance the overall benefit from such multilateral coordination (as was done, for example, with the International Halley Watch).

The IACG will continue its role as a forum for inter-agency discussions in space science for as long as deemed useful by the participating agencies. The IACG will review its overall effectiveness and continued need for existence at regular intervals.

A first point of reference is that a clear designation of a single set of approved missions in a particular area of space science is always made the focus of the IACG during a particular point in time. This goes back to the basic debate concerning the pros and cons of international cooperation generally. Some people believe that it is entirely appropriate and possible for countries to cooperate on some projects/areas while competing in others, and others believe that it is not possible to do both. As the European and Japanese space programs have matured, they have increasingly been viewed by the United States as competitive, as exemplified by the commercial launch field where the European Ariane launcher is a direct competitor to U.S. expendable launch vehicles and where the Japanese H-II is posturing to enter the market also. The Soviet Union has of course always been seen as a competitor. Traditionally, the most nonthreatening type of cooperation has been seen as that related to basic science.

With regard to the IACG and its scope, according to Michael Michaud, former director of the Office of Advanced Technology in the U.S. State Department:

> State keeps an eye on it just to see what they're up to, because it does involve the Soviets, but they have stuck to space science and as long as their agenda is consistent with the bilateral agendas that we have with those countries, then State really doesn't have a problem with it.[27]

As long as basic science is maintained as the focus of the group, other agencies will for the most part keep out. If the IACG were to become more adventurous in terms of going beyond coordination of separate projects into cooperation involving hardware or technol-ogy transfer, particularly with the Soviets, then the IACG would fall under much closer scrutiny by other agencies. In the United States, those agencies would most probably include the Department of Defense (DOD). The consequence there, of course, is that when DOD becomes involved, international cooperation often ceases to be feasible.

Beyond the political problems, focusing on coordination minimizes the potential technical problems which can be involved where there is cooperation on hardware. Quadrilateral cooperation on hardware could be a bureaucratic and managerial nightmare, as evidenced by the length and complexity of the Space Station negotiations. Even cooperative agreements of a bilateral nature are often difficult because of control issues, and on a practical level these problems would be multiplied by four within the IACG.

Another point consistently made is that the IACG is a vehicle for coordination of already approved projects, not for planning future projects. Allowing the IACG to deal only with approved projects avoids endorsements being made by IACG delegates that then cannot be followed through on, or using those endorsements as domestic levers for program approval.

The provision concerning IACG leadership is to ensure that the IACG actually has the ability to be effective in its coordination efforts. The idea was that without agency representation from the senior level, discussions might be little more than wishful rhetoric.

Finally, the Terms of Reference mandate that the IACG will coordinate projects of mutual interest to all parties. The current membership of the IACG is NASA, ESA, ISAS, and Intercosmos; the IACG is thus limited to projects where all have expressed an interest. Otherwise, the feeling is that other arrangements, perhaps of a bilateral nature, are more appropriate.

Within the United States approval for continuing the IACG, though not unopposed, was fairly straightforward. Objections were for the most part nebulous, and therefore not pursued since the Terms of Reference guidelines were unambiguous. Michael Michaud explained that:

> The thing that makes us uncomfortable is that the IACG lumps together our friends and our allies with the Soviets. We know how to conduct relations with our allies; we know how to conduct relations with the Soviet Union. But we do it differently. The problem with IACG is that it lumps everything together.[28]

In November of 1986 when the U.S. decision to continue its participation in the IACG was made, the State Department and the National Security Council made it clear that they wanted the organization kept clearly focused on space science. Earlier, while serving as acting NASA administrator, William Graham had gone so far as to send a still classified letter to NSC Director John Poindexter, expressing his reservations, but nothing came of it. Relations with the Soviets were improving, so the time was propitious. Also, NASA was still busy dealing with the *Challenger* accident and Poindexter was busy with still-unknown-to-the-public Iran-Contra concerns. As Burton Edelson later explained, since nobody was violently opposed to the group and it was basically a small matter compared to other goings on, the IACG was allowed to continue under the Terms of Reference.[29]

In comparison, acceptance and continuance of the IACG in Europe, the Soviet Union, and Japan was viewed as a positive and desirable step. The Europeans and Soviets felt that in many ways they were responsible for the existence of the group, and proud of that fact. The Japanese felt a deep sense of pride in their IACG participation, as well as in the success of their own national spacecraft. Further, for ESA and ISAS, IACG is a group in which they, while being smaller space agencies, are treated as equal partners.

The IACG in the Present

The Solar Terrestrial Science Project

Parallel to scientific preparation for the Halley's Comet encounters, other international scientists were already working on various projects concerning solar terrestrial science. Coordination and cooperation became a factor relatively early in many of those efforts.

When Burton Edelson first joined NASA in 1982, he wanted to establish a joint science planning program between ESA/NASA. Joint planning was a new concept, as in the past, international cooperation in space had consisted primarily of the United States merely helping other countries or coordinating activities with other countries, as part of an assertion of U.S. leadership. Edelson managed to get the joint planning idea included in the minutes of a June 1982 meeting between NASA Administrator James Beggs and ESA Director-General Erik Quistgaard, which Edelson too was attending in his role as associate administrator for space science and applications. The minutes were duly signed with little notice or attention given to this departure from the past.

Solar terrestrial physics was among the first areas targeted for joint planning by Edelson and his ESA counterpart, Roger Bonnet. Originally, "joint planning" had been implemented merely by holding two meetings each year, one in Europe and one in the United States, where each side made a presentation to the other as to what they were doing. Over the years, however, the presentations evolved into detailed discussions, including suggestions for the future and planning for how each side might become involved in the other's projects.

Together, Edelson and Bonnet began to put together an international project involving NASA's International Solar Terrestrial Physics (ISTP) project, and ESA's Solar Terrestrial Science Program (STSP). The resulting coordinated project was, according to Edelson, "100 percent the product of joint planning,"[30] The concept of an international solar terrestrial physics program began to emerge with the realization that to accomplish all, or nearly all, of the scientific goals set by the countries interested, a large number of spacecraft with coordinated instrumentation would be needed to do solar physics and plasma physics. A number of bilateral agreements have been made to accomplish this task: between NASA–ESA[31]; between NASA–ISAS,[32] as the United States had separate dealings with the Japanese since they had a similar program in solar terrestrial physics; and between ESA–IKI,[33] as the United States had no agreement with the Soviets until 1987.[34] From these, the IACG Solar Terrestrial Science project has evolved into a program involving approximately thirty satellites and quadrilateral coordination between the countries involved.

At the IACG meeting in Washington, three areas were identified for study concerning future coordination within the IACG framework: solar terrestrial science; planetary and primitive bodies; and space very large baseline interferometry (VLBI). In Padua in 1986, the IACG decided that solar terrestrial science should be the next area of focus for the IACG. This was considered the logical choice for two reasons: the IACG Terms of Reference clearly state that areas targeted for IACG coordination should be of mutual interest to all parties, and solar terrestrial was the only area where all four agencies had approved programs; further, work on solar terrestrial physics already had a strong background of coordinated efforts. Two working groups were initially formed for the Solar Terrestrial Science project: Working Group 1, Science, and Working Group 2, Data Exchange.

The October 1987 IACG meeting in Kyoto, Japan accordingly focused on solar terrestrial science. Reports were given by each delegation concerning their national projects. These included:[35]

- ESA: Cluster, SOHO, Ulysses
- NASA: CRESS, UARS, Polar, Wind
- Intercosmos: Interball, Relict-2, IKI-1, IKI-2
- ISAS: Geotail, EXOS-D, Solar-A

Although the IACG charter mandates that they maintain a focus on coordination of projects only, bilateral agreements have made aspects of the Solar Terrestrial Science project a cooperative project also, as they resulted in some cooperation on satellite design.

The 1988 meeting of the IACG was held in Cocoa Beach, Florida. Substantively, perhaps the most important action taken at the meeting was to establish a third working group, Mission Design and Planning, for the Solar Terrestrial Science program. Probably of equal importance though, were the changes of personnel. Dr. Lennard A. Fisk had replaced Burton Edelson at NASA after Edelson left in 1987, and thus became the U.S. delegation head to the IACG. Also, for the first time Sagdeev did not head the Soviet delegation. In fact, due to other obligations, he was unable to attend the meeting at all. Galeev assumed the delegation head position. Following Sagdeev, with his charismatic personality, would be difficult for anyone. But Galeev rose to the occasion and handled the situation well, making it clear that he intended to carry on with the spirit of cooperation set by Sagdeev, while establishing his own leadership role within the group. In addition, Professor Jun Nishimura served as the IASA delegation head for the first time, taking over from Professor Oda. Clearly, there were more new faces than at past meetings, where the IACG members had become like a group of close friends. At the beginning of the meeting, delegations stayed to themselves more than they might have one or two years before. But by the end of the five day meeting, people were feeling more comfortable with each other—a key factor of IACG success. The 1989 meeting of the IACG took place in Prague, Czechoslovakia.

Other IACG Project Interests

Although the vast majority of the IACG work is at the working group level, which deals with approved programs, the IACG also includes panels to work on potential projects for the future. So although the IACG is not focusing on solar terrestrial science, panels are still monitoring and reporting on activities in the other two areas of specified interest, space VLBI, and planetary and primitive bodies, with reports given at the annual meetings. Also in Kyoto, Cocoa Beach, and Prague there was discussion concerning the International Space Year in 1992. It was felt that the Solar Terrestrial Science project would be "an excellent example of successful international cooperation in space science and this should be fully exploited for ISY purposes."[36]

The IACG as an Organization

Three questions emerge from this examination of the beginnings of the IACG and its success with its first project, which reference the issues raised in 1986 by Pedersen. Will the IACG, like most international organizations, become more institutionalized in the future, and perhaps with that institutionalization lose some of the informality and collegiality that has been at least partially responsible for its success? Was the collegiality and informality so obvious during the Halley's encounter the result of unique personalities with common goals working together during a time of high stress and excitement, and therefore could the success of the IACG be threatened by personnel changes? Is the IACG the proper institution to coordinate a Mars mission, manned or unmanned, which is currently a topic of a great deal of discussion? Answers, of course, can only be speculative and sometimes subjective.

Concerning the prospects for institutionalization, it is highly unlikely that the IACG will become institutionalized in the sense of establishing a permanent secretariat. A secretariat, or bureaucracy, is usually established with the intent that it maintains a system or process of carrying out the day-to-day business of the organization or, in this case, the projects it supports. According to the Terms of Reference, the IACG deals only with projects already approved at the national level. Subsequently these projects are implemented and supported at the national level, and therefore would continue anyway even if the IACG were not in existence, creating no need for a secretariat. The IACG is thus desirable for enhancement of the project, but not essential to the carrying out of individual projects themselves.

Furthermore, the participants themselves recognize the desirability of keeping the organization informal. According to Reinhard in a 1986 discussion of the future of the IACG:

> All agencies agree that it is desirable to continue the IACG beyond 1986 on an informal, non-institutionalised basis, as this will allow greater flexibility in maximising the information exchanges. . . .[37]

This informal nature also maintains the nonthreatening posture of the IACG to the national governments represented by the group; the nonthreatening posture being necessary to appease other national-security-oriented government agencies and hence allow continuance of the IACG as an entity.

It is important to note, however, that non-institutionalization does not necessarily equate with nonexpansion. The membership rule for both Halley's and the Solar Terrestrial Science project has been that an agency must have a dedicated spacecraft in the field of focus. NASA inclusion on Halley's was formally through the ICE spacecraft. As other space agencies increase their capabilities, have the desire to participate in IACG activities, and meet the spacecraft criteria, the membership of the IACG could be expanded on future projects.

The question of importance of particular individuals is more difficult. At the senior level, Sagdeev, Edelson, Bonnet, and Oda were all committed internationalists. Their combined presence was perhaps essential to get the organization started, but now that the structure is in place, hopefully it will take on a life of its own. At the working group level, because the subject matter of the second phase has changed, many members and leaders of the three Solar Terrestrial Science working groups are new. There were questions, and concerns, during the transition from Phase 1—Halley's Comet to Phase 2—Solar Terrestrial Science as to whether the new people would understand and accept the unique working mode of the IACG. It is critical that they do, because it is at this level that most ideas for coordination are created and carried out.

Finally, a mission to Mars is perhaps the space science project drawing the most public attention of late. The success of the IACG and Halley's Comet has led to speculation as to whether the IACG might most appropriately organize such a mission. At this time the answer is probably no for some very simple reasons.

The IACG is to coordinate already approved projects of mutual interest to IACG members. At present, neither ESA nor Japan have any approved projects concerning Mars, or even any specific interest in that area. The Soviets have the most active Mars program, but even it is not without some questions. In the United States, the Mars observer project originally scheduled for launch in 1985 has now been pushed to at least 1992. Beyond that, there are no approved Mars projects. Therefore, a Mars mission does not fit the criteria for an IACG project. A joint unmanned Mars sample return, probably the only politically acceptable mission in the United States, would perhaps be more appropriately done as a bilateral U.S.-Soviet project.

Currently, the planetary panel of the IACG is engaged primarily in the exchange of information concerning national programs. The space VLBI panel is actually far more active and seems to be working toward being the next IACG project. Another space science area where an IACG panel might be appropriate and useful, as all four space agencies are studying the topic, concerns a solar probe. That would be both technically challenging and scientifically exciting, as the spacecraft would fly extremely close to the Sun.

Conclusion

The IACG is an organization which in its short lifetime has achieved one remarkable success. Its success can be attributed to several factors. The clear commitment of the people involved and their ability so far to circumvent bureaucratic obstacles have been essential. Also, IACG focus on space science has allowed it to avoid being consumed by the political forces which usually accompany space ventures which have any national security aspects or the potential for commercial profit.

Yet the IACG is still fragile, in the sense that the organization must continually prove itself to the perpetual skeptics about the benefits of cooperation outweighing the costs. There is always the chance that someone will get frustrated with an organizational, communication, or some other type of problem and simply decide to pull out—as the national program would continue anyway.

Surviving the transition from Phase 1 to Phase 2 was a major milestone which has stabilized the IACG to a large extent. The Solar Terrestrial Science program of the IACG will last until the end of this century; this should assure the group's existence for at least another decade.

With regard to Pedersen's proposition that the IACG may be an incremental step toward an International Space Science Agency, it may well be, but not within the near future. The success of the IACG during the Halley's Comet encounters was based on the conscious efforts of its participants to stay within careful limits of national political propriety. Expansion in the sense of giving the IACG a more independent status would in fact be a gross violation of those limits, and would require fundamental changes in attitude by at least two of the current members, the United States and the Soviet Union. In the immediate future, then, perhaps the most important general tasks of the IACG will be to show that cooperation can be successful through projects like Solar Terrestrial Science, and hence to convince those holding the "cooperate or compete but not both" position of the fallacy of that premise. Then the attitude change necessary for an ISSA to exist may be possible. In the meantime, however, the IACG can also serve as a valuable model for cooperative ventures already underway—like global remote sensing—in that the reasons it has worked are clear. Learning from successful cooperative organizations—INTELSAT, Arianespace, ESA, and the IACG among others—and then applying applicable aspects to new ventures will promote and accelerate the evolutionary process leading toward an ISSA type of organization.

Endnotes

1. An earlier version of this chapter appeared in *Space Policy*, November 1989.
2. Kenneth S. Pedersen, "The Changing Face of International Space Cooperation: One view of NASA," *Space Policy*, May 1986, p. 128.
3. National Commission on Space, *Pioneering the Space Frontier*, (New York: Bantam Books, 1986), p. 165.
4. Kenneth S. Pedersen, "The Global Context: Changes and Challenges," *Economics and Technology in U.S. Space Policy*, Molly Macauley, ed., Proceedings of a Symposium held in Washington, D.C., June 24–25, 1986, Resources for the Future and the National Academy of Engineering, p. 187.

5. Ibid.

6. Ibid.

7. For a complete chronicle of the U.S. decision, see: John M. Logsdon, "Missing Halley's Comet:" The Politics of Big Science," *Isis*, June 1989.

8. For a list of U.S. space observations of Comet Halley that were eventually carried out, see: R. Reinhard, "The Role of the Inter-Agency Consultative Group (IACG) and its Associated Working Groups," *ESA Bulletin*, May 1986, p. 83.

9. The relevance of the meeting site relates to a painting, "The Adoration of the Magi" which is part of a fresco decorating the interior of the Scrovegni Chapel in Padua. Painted in approximately 1303 by Florentine master Giotto di Bondone, it depicts the Star of Bethlehem as a comet. This representation is thought to have been inspired by Halley's Comet, which had appeared in 1301.

10. "Space Missions to Halley's Comet and Related Activities: The Inter-Agency Meeting in Padova, 13–15 September 1981," *ESA Bulletin*, February 1982, p. 65.

11. R. Reinhard, "The Role of the Inter-Agency Consultative Group (IACG) and Its Associated Working Groups" *ESA Bulletin*, May 1986, p. 84.

12. Interim meetings had been held in 1982 in Dobogoko, Hungary: 1983 at Kagoshima, Japan; and 1984 at Talinn, in the Soviet Union.

13. *The Halley Armada*, NASA, Jet Propulsion Laboratory, JPL 400–278, Rev. 2, 10/85.

14. Interview with Rudegar Reinhard, Paris, 7 June 1988.

15. Interview with Rudegar Reinhard, Paris, 7 June 1988.

16. See: Nicholas Daniloff, "The Space Statesman," *Air & Space*, October/November 1988, pp. 42–46.

17. This third group was abolished in 1983 as unnecessary, because all the relevant tasks had been incorporated into Groups 1 and 2.

18. With the exception of a few representatives who were in Moscow and Darmstadt, the Japanese remained at their own facilities in Japan for their flybys.

19. For a detailed description of the Pathfinder concept, see: R. E. Munch, "The Pathfinder Concept, Its Implementation and Its Results," *ESA Bulletin*, May 1986, pp. 75–79.

20. Giotto was a particular source of pride for ESA, as it was the first truly European large-scale interplanetary space science mission. The Giotto craft was designed and built by a consortium of European space industries, with British Aerospace acting as prime contractor.

21. For a table of key data on missions to Halley's Comet, see: Reinhard, "The Role of the Inter-Agency Consultative Group (IACG) and its Associated Working Groups," p. 81.

22. Interview with R.E. Munch, Darmstadt, 13 June 1988.

23. See: R. Reinhard, "Comet Halley—After the Heidelberg Symposium," *ESA Bulletin*, February 1987, pp. 62–75.

24. After the success of the Halley's armada, elaborate arrangements were made for the delegation heads to go to Rome after the business meeting in Padua, to make presentations to the President of Italy and to Pope John Paul II about the success of the missions, and the IACG. *Encounter '86, An International Rendezvous with Halley's Comet*, BR-27, ESA Publications Division, November 1986.

25. Interview with R. Reinhard, Paris, 7 June 1988.

26. Inter-Agency Consultative Group for Space Science (IACG), Terms of Reference, p. 1.

27. Interview with Michael Michaud, Washington, D.C., 4 May 1988.

28. Interview with Michael Michaud, Washington, D.C., 4 May 1988.

29. Interview with Dr. Burton Edelson, Washington, D.C., May 6, 1988.

30. Interview with Dr. Burton Edelson, Washington, D.C., May 6, 1988.

31. The NASA-ESA agreements concern the Solar Terrestrial Science Program.

32. A soft x-ray telescope to be flown on the ISAS Solar-A is a joint NASA/ISAS experiment. Geotail, a high altitude satellite with an orbit maneuver capability, will be launched by NASA using a Delta II expendable launch vehicle in summer 1992. It will also carry NASA experiments.

33. There is an Intercosmos/ESA assessment study being done on two new Soviet satellites similar to the ESA Cluster spacecraft.

34. Note: the U.S.-Soviet cooperative agreement signed in the Spring of 1987 is for coordination of projects only, so it would still not involve cooperation on hardware.

35. For a complete description, see: *Summary Report*, Seventh Meeting of the IACG for Space Science, Kyoto, 20–21 October 1987, sections 2.1–2.4.

36. Ibid., section 6.

37. Reinhard, "The Role of the Inter-Agency Consultative Group (IACG) and its Associated Working Groups," p. 94.

Chapter 16

Conclusions

Because the United States began as the dominant space power concerning cooperative ventures, it has never had to learn to operate in any manner other than "the U.S. way." But things have changed. There are now an increasing number of space "actors" with varying ranges of capabilities, including the Soviet Union, the Europeans collectively and in some cases individually, Japan, China, Canada, and many third world countries. The United States is no longer "the only game in town" in space activities, although in some cases it is still trying to act as though it is. These other countries have worked hard to establish a place for themselves in the international space arena, and along the way they have learned some valuable lessons.

Clearly, roles and relationships in space are being redefined—internationally, nationally, and locally in some cases. People and groups will have to learn to work together, to cooperate, as some have not had to in the past. All of which, experience shows, leads toward the goal of increased space related activity. Some clear and interrelated propositions can be extracted from history, in terms of lessons to be learned, that will maximize opportunities for successful cooperative ventures in the future.

Cooperation is Not a Luxury, It is a Means of Survival

Within the context of the European Space Agency, the Europeans learned early that if they did not work together, sharing both their knowledge and their capital, they would fall so far behind the United States technologically that they would never be able to catch up. So they have worked together and have provided a model for other groups that would like to do the same. The Canadians have also shown that cooperation can be an extremely efficient way of operating. The need to work together becomes even more imperative in the face of ever-tightening budgets, strengthening this proposition even further.

Technology is the Key to a Strong Economic Future

Under present economic conditions, no country, group of countries, state, or private sector organization has money to spend on a frivolous venture—as some critics choose to characterize space. We must question, then, why many nations are prepared to invest so heavily in space ventures; they apparently have realized something the United States has not. The returns from investment in space—both short-term and long-term—are going to play a major role in future national economic pictures. Most other space-faring nations have already acknowledged the need for a close relationship between industry and government where the space program is concerned. Investment by Japanese aerospace industries in space programs is not for esoteric or altruistic reasons. The commitment of China and India to the development of an active space program is further evidence of the importance other countries are giving to space in their future. Yet it is only recently that U.S. policy makers have shown any cognizance of the role that space will most certainly play in determining future industrial growth and leadership in this country. If the private and public sectors in the United States worked closer together to support technological advancement, it would certainly make it easier for congressional representatives to support space programs in the budget. The Europeans and Japanese recognized this long ago, while we in the United States have been slow to acknowledge the need to adapt our institutions and ways of doing things in the face of an increasingly cooperative and competitive globalization of space activities.

Cooperation in Space Should Not Be Expected to Result in a Generally Improved Political Environment

Cooperation in space has traditionally been an outgrowth of good political relations. There is little evidence that the relationship works the other way to any significant degree and to expect such, especially in the short term, is unrealistic and a scenario for failure. The value of cooperation in space to the more general political climate comes through the willingness shown by the involved countries to try to work together and, in the case of adversaries, to reduce tension and achieve common goals.

In an environment of increasingly expanding players, the ability of a country to actively participate in cooperative ventures will have significant leadership implications

Leadership is not, and should not, be defined as the country playing the biggest role in every field of space activity or every space project. Even under the best of economic situations, which is certainly not the case under Gramm-Rudman-Hollings federal budget cuts, it would be impossible for the United States to take up singlehandedly every challenge presented by space or even to be the dominant player in every cooperative venture. Therefore more cooperation can increase opportunities for involvement and investigation in a variety of fields, with varying levels of participation.

The whole issue of leadership also encompasses another aspect—that nowhere in U.S. policy is "leadership" specifically defined. Yet it is taken for granted that "leadership" should be a driver for policy planning. Perhaps "leadership" should more appropriately be viewed as a result of policy. In Europe, for example, the policy goal has been stated as autonomy in space. If current long-term plans are carried through to a significant degree, it is highly probable that this policy will result in Europe taking a leading role in certain areas of space activity. The same was true for the United States under the Apollo program. Now we have set leadership itself up as both a reason for, and a goal of, space activity. The United States needs to put leadership in a more defined perspective.

Certainly one aspect of leadership is the ability to guide, usually through exemplary conduct and advanced capabilities. The United States has and continues to (for now) enjoy leadership in most areas of space technologies. It is sadly lacking, however, in political commitment, as evidenced in its inability to follow through with an aggressive, well-planned agenda.

The Soviet Union received a significant boost in prestige as a result of its activities and the cooperative attitude exhibited during the Halley's encounters. This was in contrast with the low-keyed role taken by the United States and the subsequent lack of public relations benefits reaped by the United States. Clearly, this evidences the marked impact that involvement in cooperative ventures has on perceived leadership ability. Questions about whether the Soviets are challeng-

ing the United States for the leadership role in space are not the result of quantum leaps in Soviet technology over the past four or five years, but rather due to their new willingness to step in and make offers for participation on cooperative ventures as the United States did in the 1960s, 1970s, and early 1980s. As the number of countries with mature space programs increases, the necessity of maintaining a dynamic space program in order to positively influence the activities of other countries to complement your own program will increase also.

International cooperation in space will occur in areas of non-national security oriented, "low-politics" fields

In this respect space has much in common with more traditional areas of politics: chances of successful cooperation, success here being defined from a political perspective rather than a scientific one, increase as the national security aspects of issues involved decrease. That is, successful cooperation is easier in international mail delivery than in arms control. Within political science, the functionalist and neofunctionalist theories have supported this premise since the late 1950s. Functionalism states that cooperation will inherently begin in areas of low politics and then spill over to other areas after a certain degree of trust and necessity warrants such. Neofunctionalism takes that premise one step further and, simply stated, says that even low level cooperation will begin at a regional level. The experience of ESA as well as the new regional cooperative mechanisms within the third world certainly seems to support this neo-functionalist premise.

Within the area of space politics specifically, the functionalist premises of cooperation clearly hold true. The experiences of the IACG with its deliberate focus on space science evidence the need to keep activities within this perceived low politics realm in order to increase their chances of success. The U.S. plan regarding development of a hypersonic vehicle, the national aerospace plane (NASP) or the X-30, is an example of projects with national security implications being considered inappropriate for cooperative efforts.

The X-30 is an experimental technology R&D program with several potential applications. Use for civilian transport, the Reagan administration's Orient Express, is probably furthest down the road for development. More immediate applications are as a vehicle which will significantly decrease the cost of putting payloads into space, thereby making space truly accessible on a commercial basis, and as a military vehicle. The program was started by the Defense Advanced Research Projects Agency (DARPA) and then transferred to the Air Force. The funding of the project is an 80–20 percent split between DOD and NASA. Development costs are estimated at $3.3 billion between 1986 and 1994. Obviously an R&D program of this size needs congressional support to be sustained. Yet it has been a totally black military program, putting it in constant jeopardy as a target for the budget ax—well illustrated in 1989 when the program was almost cut. Although France, Japan, West Germany, and the Soviets are working on hypersonic designs also, collaboration even on basic R&D to decrease costs is unlikely. A 1988 General Accounting Office study concluded that the United States "has no plans for foreign participation in developing the X-30" as "political, economic, financial, technological and legal reasons make international cooperation in developing the X-30 undesirable."[1] As a research program which many people feel encompasses a significant number of technologies critical to the next century, both economically and politically, continued U.S. involvement with hypersonic research also has important technological leadership implications.

In the future, it is highly probable that areas of low threat and maximum return of benefits will continue to be the ones where the most cooperative work will be undertaken and prove successful. It is interesting to note, however, that there has even been reluctance to work together on issues which upon first examination seem inherently appropriate for cooperative action. Space debris is a problem which truly has no national boundaries, has aspects of being both an international environmental and legal issue. It does not carry obvious national security implications. Yet although it is potentially an area appropriate for cooperative remedial action, that has not entirely been the case to date.

> At its plenary meetings in 1984 and 1986, the Committee on Space Research (COSPAR) . . . held workshops devoted exclusively to the topic of space debris. Although the majority of participants were from the United States, papers from Czechoslovakia, the European Space Operations Center, and West Germany indicated the awakening recognition of the space debris problem around the world. Notably absent from international forums on space debris has been the Soviet Union.[2]

The reluctance of the Soviet Union to become involved probably stems from the same reason as that causing reluctance among some satellite owners and operators—the very group most likely to be affected by the problem—which is fear that the remedy will impinge upon their activities. Any measures which restrict the use of space can be interpreted as affecting national interest. So the issue of space debris may be neglected until it begins to affect space operations detrimentally.

Subjecting space programs to an annual budget process ends up costing more money in the end and is a very inefficient process

Along with maintaining a dynamic space program, actors must be able to follow through on the plans. Trust develops with positive experiences, and repeated impediments to positive experiences must be dealt with. In the United States, that will mean dealing with the budget situation. This situation is well recognized, as evidenced from the 1987 report on the U.S. space program by the American Institute of Aeronautics and Astronautics which says:

> A necessary ingredient for improved cooperation requires that some mechanism be created for committing budgets over several years so that we would not be forced to renege on our promises.[3]

Although the ISPM debacle was the exception to the rule, the memory of it will not soon be forgotten, and with just cause. Clearly, under current circumstances the United States cannot embark upon a space project with any degree of certainty that it will be carried through to completion. The havoc that this creates for a purely domestic project is substantial, for cooperative projects it is potentially prohibitory.

Also within the context of space budget debates, the notion that "If you don't do X, then you can do Y" needs to be examined and better understood. Part of the rationale for making priority choices through annual budgeting, this argument states, is that if money were not spent on space projects, then we would have more money for better roads, health care, education, and so on. It was also part of the argument made during ASTP by critics who felt the money could have been better used elsewhere. Buying into this line of thought is easy, and it certainly is used by politicians when convenient and likely to make points for them. Yet probably it is based on false assumptions. Certainly money not spent in one area—like space—makes the overall amount of available money for other programs larger. But to assume that the money would then be authorized and appropriated for education, housing, or social service programs is rather a quantum leap. Money taken from the space budget—where there is a return of about $7 into the economy from each $1 invested—would most likely be dispersed in a manner highly unlikely to directly benefit the general public. Yet there is an assumption that such would be the case. The complexity of the budget process—as well as the traditional assumption that without an advanced degree one is relegated to being an observer of the space program and cannot possibly understand what is going on—has made arguments like these easy for the politicians to make and easy for the general public to accept.

Let criticism and disagreement play a constructive role

It is important to recognize that although the Europeans present a relatively united front in negotiations outside of Europe, they certainly have their internal difficulties too. In Europe, however, disagreements are far more likely to stay behind closed doors than they are in the United States. Americans are more open about strains within the system, sometimes raising speculations about the certainty of a project or policy and consequently casting additional doubt on the already strained worldwide view of U.S. political commitment. What often seems to happen in the United States is that at closed door meetings, politicians are so anxious to have the appearance of cooperation that valid disagreements are suppressed. This results in disgruntled individuals turning what might have been constructive criticism into useless frustration and venting it beyond the closed doors and probably within earshot of the press. On the other hand, constructive criticism offered publicly is too often seen as challenging the space program, and attempts are made to surpress it. No program, U.S., Soviet, European, Chinese, or whatever, is beyond improvement. Those individuals most threatened by constructive criticism are often the ones most parochial in both interests and outlook, and hence part of the problem themselves. The key concerning criticism is to differentiate between frustration, which often can be prevented; criticism for the sake of criticism, which is often the sensational variety which the media should resist the temptation to indulge in or encourage with undue attention; and constructive criticism, often identifiable in that it not only points out problems but offers suggestions as to how they might be corrected.

Tying long-term goals and programs, both cooperative and national, to short-term projects will make them more palatable and sustainable

Incrementalism is a word unknown in the U.S. space program, although it is the norm elsewhere. The Soviets for many years were thought of as plodding in their space ventures because they used the same rocket designs, the same space station designs, and the same satellite designs year after year with only small modifications. Meanwhile the United States was developing the Saturn V and discarding it, creating Skylab and abandoning it, designing a Space Station, and having to fight tooth and nail not to have it scrapped by Congress. It is interesting to note that the Soviet space program began to run into problems when it began to imitate the U.S. "go for the spectacular" attitude, especially concerning Buran. The point is, incremental progress is not bad, as it results in steady progress, something that the United States could do well to have more of. Certainly the cost benefits of working incrementally are obvious.

The entire debate surrounding a possible manned mission to Mars, whether as a joint venture with international partners or as a U.S. venture alone, involves several aspects of this point. Should the United States plan a manned mission to Mars? Should it be modeled after the Apollo program, as the fastest way? Or should we proceed more along an incremental plan like that suggested by the National Commission on Space[4] in 1986, and the 1987 report of the commission led by former astronaut Sally Ride, entitled LEADERSHIP AND AMERICA'S FUTURE IN SPACE, both of which call for the creation of a space infrastructure along the way for repeated and extended voyages as part of the plan?[5] If either of those scenarios is approved, should it be strictly a national program or should it be a joint program with the Soviets? Or should it be a multinational program with the Soviets, Europeans, Japanese, and any others who could participate in some meaningful way?

As an international program of whatever configuration of countries, probably the only way a manned Mars mission would have any chance of success is an incremental program. Since neither the Europeans nor the Japanese have extensive plans involving Mars at this time, a joint U.S.-Soviet configuration becomes the most likely, beginning perhaps with a joint unmanned program to send a lander and rover to the Martian surface. That scenario has been suggested repeatedly by Dr. Roald Sagdeev and others.[6] It would minimize the political risks involved with committing to a long-term program between basically adversarial countries, minimize the technology-transfer problems since each country would build unique equipment, and minimize the expenses since costs would be shared and spread out.

The Soviets have had the most vigorous Mars program of late, although those too have been scaled back recently. The U.S. has been hesitant, primarily for political and economic reasons, but willing to listen to proposals for joint ventures to Mars with the Soviets. Unfortunately, U.S. hesitancy may increase as a result of the Soviets losing control of both of their Phobos spacecraft. Now doubts about technical competence may be added to the list of doubts about the wisdom of a cooperative mission. Clearly though, if there is any chance for a cooperative Mars mission, it will have to as part of an incremental program.

The prospects for international cooperation on space ventures in the future are promising. Recognition that it is possible to compete and cooperate on differing areas of space activity simultaneously is critical. Subsequently, identification of particular areas where co-operation can be mutually beneficial for all parties will increase the potential success of the venture. Most importantly, however, increased cooperation will increase the opportunities for activity in space. Although often presented as the final frontier, space is certainly the next frontier, one the United States cannot afford to neglect or take lightly.

Endnotes

1. U.S. General Accounting Office, *National Aero-Space Plane: A Technology Development and Demonstration Program to Build the X-30*, GAO/NSIAD-88-122. April 1988, p. 48.

2. Nicholas L. Johnson and Darren S. McKnight, *Artificial Space Debris* (Malibar, Fla.: Orbit Book Co., 1987), p. 87.

3. "U.S. Civil Space Program: An AIAA Assessment," American Institute of Aeronautics and Astronautics, March 1987, p. 20.

4. *Pioneering the Space Frontier*, The Report of the National Commission on Space (New York: Bantam Books, 1986).

5. See: Isaac Asimov, "Lost in Space," *Discover*, January 1988, pp. 18–20; John Logsdon, "Resist the Pull of Mars," *Air & Space*, April/May 1988, pp. 44–45.

6. For discussions of what might be involved with an international mission to Mars under various scenarios see: Michael A. G. Michaud, "Planetary Partners," *Ad Astra*, March 1989, pp. 25–28; Burton I. Edelson and John McLucas, "U.S. and Soviet Planetary Exploration: The Next Step Is Mars, Together," *Space Policy*, November 1988, pp. 347–349; Burton I. Edelson and Alan Townsend, "U.S.–Soviet Cooperation: Opportunities in Space," *SAIS Review*, The Johns Hopkins University, Winter–Spring 1989, pp. 183–197.

Index